DEVELOPMENTAL ASPECTS OF CARCINOGENESIS AND IMMUNITY

The Thirty-Second Symposium of
The Society for Developmental Biology

Manhattan, Kansas, June 13-15, 1973

SPONSORED BY
THE SOCIETY FOR DEVELOPMENTAL BIOLOGY, INC.
and
SUPPORTED IN PART BY
THE NATIONAL CANCER INSTITUTE*
NATIONAL INSTITUTES OF HEALTH
DEPARTMENT OF HEALTH, EDUCATION, AND WELFARE
BETHESDA, MARYLAND, U.S.A.

*DIVISION OF CANCER RESEARCH RESOURCES AND CENTERS,
GRANT NO. 1 R13 CA-14512-01

Developmental Aspects of Carcinogenesis and Immunity

Edited by

Thomas J. King

Division of Cancer Research Resources and Centers
National Cancer Institute

1974

ACADEMIC PRESS, INC. *New York   San Francisco   London*

A Subsidiary of Harcourt Brace Jovanovich, Publishers

ACADEMIC PRESS, INC.
111 Fifth Avenue, New York, New York 10003

United Kingdom Edition published by
ACADEMIC PRESS, INC. (LONDON) LTD.
24/28 Oval Road, London NW1

LIBRARY OF CONGRESS CATALOG CARD NUMBER: 55-10678

ISBN 0–12–612977–0

PRINTED IN THE UNITED STATES OF AMERICA

Contents

CONTENTS

List of Contributors

Bold Face Denotes Session Moderators

Norman G. Anderson, The Molecular Anatomy (MAN) Program, Oak Ridge National Laboratory, Oak Ridge, Tennessee 37830

Renato Baserga, Department of Pathology and Fels Research Institute, Temple University School of Medicine, Philadelphia, Pennsylvania 19140

Boyce W. Burge, Department of Biology, Massachusetts Institute of Technology, Cambridge, Massachusetts 02139

LaRoy N. Castor, The Institute for Cancer Research, Fox Chase Center for Cancer and Medical Sciences, Philadelphia, Pennsylvania 19111

Joseph H. Coggin, Jr., The University of Tennessee, Department of Microbiology, Knoxville, Tennessee 37916

Michael Edidin, Department of Biology, The Johns Hopkins University, Charles and 34th Streets, Baltimore, Maryland 21218

Chil Yong-Kang, McArdle Laboratory, University of Wisconsin-Madison, Madison, Wisconsin 53706

Jung-Chung Lin, Department of Pathology and Fels Research Institute, Temple University School of Medicine, Philadelphia, Pennsylvania 19140

Frederick Meins, Jr., Department of Biology, Princeton University, Princeton, New Jersey 08540

G.J.V. Nossal, The Walter and Eliza Hall, Institute of Medical Research, Melbourne, Victoria 3050, Australia

G. Barry Pierce, Department of Pathology, University of Colorado **Medical Center, Denver, Colorado 80220**

Henry C. Pitot, McArdle Laboratory for Cancer Research, University of Wisconsin School of Medicine, Madison, Wisconsin 53706

Sylvia Pollack, Department of Microbiology, University of Washington Medical School, Seattle, Washington 98195

Richmond T. Prehn, The Institute for Cancer Research, 7701 Burholme Avenue, Fox Chase, Philadelphia, Pennsylvania 19111

Martin C. Raff, Medical Research Council Neuroimmunology Project, Zoology Department, University College London, London WC1E 6BT

Fred Rapp, Department of Microbiology, College of Medicine, The Milton S. Hershey Medical Center of the Pennsylvania State University, Hershey, Pennsylvania 17033

Murray Rosenberg, Professor of Biology, Institute of Technology, Department of Chemical Engineering, University of Minnesota, Minneapolis, Minnesota 55455

Harry Rubin, Virus Laboratory, University of California, Berkeley, California 94720

H. Sakiyama, Department of Biology, Massachusetts Institute of Technology, Cambridge, Massachusetts 02139*

Leroy C. Stevens, The Jackson Laboratory, Bar Harbor, Maine 04609

Brian Spooner, Division of Biology, Kansas State University, Manhatten, Kansas 66506

Howard M. Temin, McArdle Laboratory, University of Wisconsin-Madison, Madison, Wisconsin 53706

George J. Todaro, National Cancer Institute, Bethesda, Maryland 20014

Gary Wickus, Department of Biology, Massachusetts Institute of Technology, Cambridge, Massachusetts 02139

*Present address: Division of Physiology and Pathology
National Institute of Radiation Science
4 Anagawa, Chiba-Shi
Chiba, Japan

Preface

The 32nd Symposium of the Society for Developmental Biology was held at Kansas State University, Manhattan, Kansas, June 13-15, 1973. In the informal atmosphere that has become the hallmark of the "Growth Symposium," the primary aim of the meeting was to bring into critical focus recent advances in carcinogenesis and immunity and emphasize their relationship to fundamental processes of developmental biology. Fourteen leading investigators presented aspects of these two rapidly moving fields that are of topical interest to students of development. In the first session G. Barry Pierce considered teratocarcinomas, and other embryonal tumors, as models of a developmental concept of cancer, while Frederick Meins made it clear that our understanding of the basic mechanisms that underlie tumor transformation must accomodate the phenomenon of potential reversibility. The second session dealt with an assessment of the voluminous data that cell contact and cell proliferation studies of neoplasia have generated, these were presented by LaRoy Castor and Renato Baserga respectively. In the same session, Henry Pitot presented evidence for considering neoplasia as a translational function of the differentiated state. The mobility of protein components of cell membranes was discussed by Michael Edidin and Boyce Burge in Session III. Fred Rapp, Howard Temin and George Todaro recounted our present knowledge of virus-mediated transformation of cells in culture and related it to developmental phenomena. The last session, on "Immunity and Oncogenesis," contained discussions of the role of embryonic and fetal antigens in cancer, lymphocyte differentiation, cell-mediated destruction of tumor cells and the enhancement of the effects of antibodies through the process of immune surveillance. Joseph Coggin, Martin Raff, Sylvia Pollack and G. J. V. Nossal made these presentations.

Nineteen Developmental Workshops, on subjects of current interest to members of the Society, were held on the first and second evenings of the meeting.

A panel discussion, "Prospects for Federal Support of Biomedical Research," was held on the afternoon of the last day. The panelists included: Thomas J. Kennedy, Jr., M.D., Associate Director for Program Planning and Evaluation, Office of the Director, National Institutes of Health; Eloise E. Clark, Ph.D., Director, Division of Biological and Medical Sciences, National Science Foundation, and Irving J. Lewis, M.A., Professor of Community Health, Albert Einstein College of Medicine.

This volume could not have been assembled without the consideration and cooperation of the invited contributors and the editorial assistance of Mrs. Lillian Davis.

THOMAS J. KING, Ph.D.

Michael Edidin   Brian Spooner   Mary Spooner   Leroy Stevens

Elizabeth Hay   Thomas F. Linsenmayer

Martin Raff   James Weston   G. J. V. Nossal

LaRoy Castor   Henry Pitot   Renato Baserga

Ursula Abbott   Anne Schauer   E. W. Hanly

Alex Haggis   John Papaconstantinou

Judson D. Sheridan   W. Sue Badman   L. Evans Roth   Winifred Doane

Frederick Meins   Leroy Stevens

Gregory S. Whitt   Prentiss Cox   Jane Westfall   Arthur Whiteley

Richmond Prehn   Sylvia Pollack

Thomas King   Thomas Kennedy   Eloise Clark   Irving Lewis

Developmental Aspects
of Carcinogenesis
and Immunity

SESSION I
Multipotentiality of the Tumorous State

Moderator: Leroy C. Stevens

THE BENIGN CELLS
OF MALIGNANT TUMORS*

G. Barry Pierce

Department of Pathology
University of Colorado Medical Center
Denver, Colorado 80220

With few exceptions pathologists are not trained in developmental biology and direct their efforts at understanding how diseases affect cells, tissues and organisms. Developmental biologists are not usually trained in pathology and direct their efforts at understanding the normal and abnormal development of tissues, organs and organisms. One would have thought that pathologists and developmental biologists would have joined forces in the study of neoplasia. On the one hand, neoplasms usually cause disease (in our lifetime more than 50 million people will be treated for cancer in this country), and on the other hand, the neoplasm is a tissue. Since all other tissues of the body originate by differentiation, one might have anticipated significant input into neoplasia by developmental biologists.

Over a half century ago, Boveri (1912), impressed by the appearance of abnormal mitosis in neoplastic tissue, proposed that the mechanism of neoplasia was a somatic mutation. It is of interest that with all of our sophisticated inputs, this idea has been neither proved nor disproved. Yet it has had a profound affect upon cancer research. Somatic mutation would explain the stability and heritability of the neoplastic change, and although it is known that chemical carcinogens can combine with DNA, RNA, and protein, the thrust has been to prove the interaction with DNA, or in other words, mutation.

The observation that cells could be made neoplastic by the introduction of a DNA virus into the genome has been construed as further support for a mutational event as the cause of cancer, but one of the important lessons emerging from studies in viral carcinogenesis is that animals in nature don't get "viral induced" tumors the way they are produced in the laboratory

*This work was supported in part by Grant ET-IN from the American Cancer Society and Grant AM-15663 from the National Institutes of Health.

(Rowe, 1973). This is not to imply that viruses can't cause cancer and that the laboratory models that have been set up are not useful tools to study carcinogenesis. What is at issue is, Where does the virus come from? Is the virus present in most cells and activated by certain environmental agents (Todaro and Huebner, 1972)? If true, carcinogenesis becomes a problem of gene or viral activation, and according to current dogma, this is the mechanism of differentiation. Differentiation is a stable process. It is a heritable process. All tissues of the body develop by differentiation — is it not surprising that systematic developmental studies in neoplasia have not been made? The explanation is multifactorial, it has to do with parochial disciplines, it has to do with our preoccupation with neoplasia as a disease, and it has to do with semantics, the expectation that the process of differentiation should lead to the production of specific gene products.

In 1967, I reviewed experiments in teratocarcinomas that had convinced me of the worthiness of a developmental concept of neoplasia. In that review is the following statement: "Although conducted upon only one kind of cancer, the experiments to be reviewed here disprove the idea of irreversibility of the malignant change, cast serious doubt upon the validity of the argument for somatic mutation as the cause of cancer, strongly suggests that the mechanisms of normal cell differentiation and stabilization apply to the pathological differentiation and stabilization of cells when they become cancerous, and strongly suggests that concepts of developmental biology will provide new approaches to therapy for cancer" (Pierce, 1967). Much has transpired in the ensuing six years. There has been maturation of thought, but more importantly, there have been many experiments which reinforce the observations upon which the developmental concept was based. Although they do not disprove the notion of somatic mutation, it turns out that there are better explanations for carcinogenesis than somatic mutation.

The premise for a developmental concept is based upon the following observations: Teratocarcinomas are known to originate in primordial germ cells, the ultimate in undifferentiated, yet multipotential cells (Stevens, 1962, 1964, 1967). After carcinogenesis, this cell evolves a caricature of embryogenesis that differs from the normal in the degree of perfection of differentiation, proliferation and organization of tissues (Pierce and Dixon, 1959; Stevens, 1960). It parallels the normal in that the result of the process is production of senescent postmitotic functional cells (Pierce et al., 1960). While many pathologists have believed that differentiation is a part of the neoplastic process, it has come as a surprise to discover that the process can yield benign cells incapable of forming tumors. These cells may even be in the majority in certain highly malignant tumors.

It is one thing to observe that differentiation occurs continually in the formation of the neoplastic mass and another to propose that the mechanism of carcinogenesis was or could or should be equated with differentiation. In other words, the notion that neoplastic tissue probably develops in the same manner as any other tissue has not been obvious, because tumors appear undifferentiated. Lack of differentiated features has been equated with loss of differentiated features, ergo dedifferentiation. Dedifferentiation is a process believed set in motion by a mutation in a differentiated cell that results in loss of differentiation and the development of the neoplastic phenotype. According to that concept, depending upon the degree of dedifferentiation would depend the degree of malignancy of the tumor. Although there is abundant evidence that cells may lose differentiated features, this type of loss has never been equated with a gain in potential – an essential prerequisite for the concept of dedifferentiation. Markert (1968) has taken the opposite point of view and views neoplasia as a disease of differentiation.

Finally, if neoplasms differ from the normal in degrees of proliferation, differentiation and organization, and since there is evidence to believe that some of these attributes are responsive to environmental control, it behooves us to determine as alternative to cytotoxic chemotherapy the means for altering gene expression and for converting malignant to benign cells (Pierce, 1967; Pierce and Johnson, 1971).

In this review I will make no attempt to be comprehensive. Although the studies on teratocarcinoma have been reviewed previously, because they are pivotal in the development of the concept and the most completely studied system, they will be briefly re-reviewed. In addition, experiments on squamous cell carcinomas, adenocarcinomas and breast carcinomas supportive of the notion will be discussed to demonstrate that the principles hold for a variety of tumors, even those of viral origin.

Teratocarcinomas

Teratocarcinomas are highly malignant tumors found most commonly in the gonads and composed of multiple kinds of somatic tissues representing each of the three embryonic germinal layers. These are often arranged in grossly recognizable organs, hence the term – teratoma (tumor of an abnormal fetus). Intermixed with the differentiated tissues is a highly malignant cancer named embryonal carcinoma because of its resemblance to primitive embryonic epithelium. Embryonal carcinoma, primitive endoderm and mesenchyme are sometimes arranged in microscopic patterns which have been called embryoid bodies because they resemble early stages in embryogenesis (Peyron, 1939; Melicow, 1940). Dixon and Moore (1953) postulated that embryonal carcinoma cells were of germ cell origin and multipotential, originating the somatic tissue by differentiation. This was the problem to which I addressed myself when I first began studying human testicular tumors. Through heterotransplantation of human embryonal carcinomas, to cortisone-treated hamsters, it was possible to obtain evidence supportive of the idea that embryonal carcinoma could differentiate into first malignant cytotrophoblast which in turn differentiated into syncytiotrophoblast forming typical chorio-carcinoma, a tumor of placental origin (Pierce and Midgley, 1963; Pierce *et al.*, 1958, 1959; Verney and Pierce, 1959). Because of difficulties in manipulating human tumors in heterotransplanted hosts, transplantable testicular teratocarcinomas of strain 129 mice were obtained from Dr. L. C. Stevens. These tumors, first described by Stevens and Little in 1954, are precise models of the tumor in man (Fig. 1, 2, 3).

Three important observations have been obtained through the study of testicular teratocarcinomas in mice. In an elegant series of experiments, Stevens identified the primordial germ cell as the cell of origin of these tumors. Since the tumors were present at birth in many instances, he studied earlier and earlier stages in their development and observed that the smallest apparently took origin in primordial germ cells of 12-day-old fetuses (Stevens, 1962). If the genital ridges of these fetuses were transplanted into the testes of adult strain 129 animals, 80% of the transplants that differentiated into testes contained small foci of neoplastic cells that ultimately developed into gross teratocarcinomas (Stevens, 1964). If a gene for absence of germ cells was incorporated into the strain 129 genome, no teratocarcinomas developed upon transplantation of genital ridges (Stevens, 1967). Subsequently, teratocarcino-genesis from primordial germ cells was studied with the electron microscope and the primordial germ cells were shown to bear close resemblance to embryonal carcinoma cells (Pierce, *et al.*, 1967) (Fig. 4).

These observations contribute importantly to our understanding of carci-nogenesis. The accepted dogma concerning carcinogenesis has been that differentiated cells become neoplastic by losing their differentiated attributes by the process of dedifferentiation. Identification of the primordial germ cell as the target, which is converted to an embryonal carcinoma cell by teratocarcinogenesis, might suggest that tumors appear undifferentiated because they have never acquired differentiated features. In other words, there is now an alternative to the notion of dedifferentiation to explain the appearance of tumors.

The second important point derived from study of this system was the demonstration by Kleinsmith and Pierce (1964) of the multipotential nature of embryonal carcinoma. About 10% of single embryonal carcinoma cells isolated in capillary pipettes, when transferred to the intraperitoneum of strain 129 mice, developed into teratocarcinomas that contained 12 or more tissues including bone, brain, glands, muscle, etc. This was the most dramatic evidence to the effect that cancer cells could differentiate. Apparently the net balance of proliferation and differentiation determines whether or not a tumor will be undifferentiated or differentiated. Again, this notion is the antithesis of dedifferentiation.

The third important point had to do with establishing that the cells and tissues derived from embryonal carcinoma were benign. It was necessary to isolate tissues derived from embryonal carcinoma to test their potential. This was done through selection, transplantation and developmental studies of embryoid bodies. It turned out that embryoid bodies could be mass produced in the intraperitoneum of mice (Pierce and Dixon, 1959). They had a life history, originated as small aggregates of embryonal carcinoma embedded in

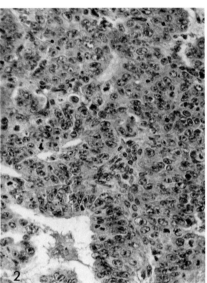

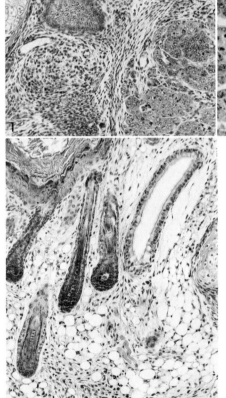

Fig. 1. *This teratocarcinoma, grown subcutaneously in strain 129 mice, is composed of elements derived from each of the 3 embryonic germinal layers. Note mesenchyme, cartilage, squamous epithelium and glands. Embryonal carcinoma may be seen at the lower right. (X 140)*

Fig. 2. *These are typical embryonal carcinoma cells. They are pleomorphic with huge nuclei contained in what appears at this level of resolution to be a syncytium of amphophylic cytoplasm. Nucleoli are prominent. Primitive neuroepithelium is differentiating at the lower left. (X 280)*

Fig. 3. *This is part of an epidermal cyst found in a subcutaneous transplant of the teratocarcinoma. Note the keratinizing squamous epithelium with hair follicles and glands. This well formed organ is distorted by a gland lined by ciliated epithelium and fragment of membrane bone (upper right). (X 140)*

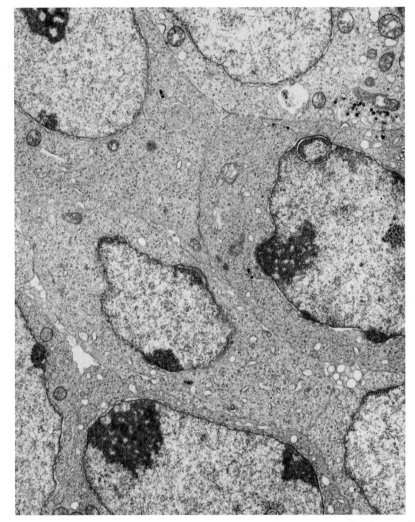

Fig. 4. *This is an electron micrograph of embryonal carcinoma cells, the rapidly proliferating stem cells of teratocarcinoma. Note the undifferentiated nature of the cytoplasm dominated by numerous polysomes. Rough endoplasmic reticulum and Golgi complexes are sparce. This is about the same degree of differentiation seen in primordial germ cells. A few viral particles are present but tumors have never been induced with cell free extracts of this tumor. (X 20,000).*

necrotic tumor tissue and were overlain by a single layer of endoderm (Pierce *et al.*, 1960). These sloughed into the peritoneum cavity, where many remained as small aggregates resembling a 5½- or 6-day-old mouse embryo. In others, mesenchymogenesis took place with migration of mesenchymal cells from the embryonal carcinoma with the ultimate formation of capillary sinusoids with hematopoietic cells. These structures became cystic and some grew to a diameter of 6 or 8 mm or more. The natural history of these embryoid bodies was studied *in vitro* (Pierce and Verney, 1961) to ensure that their development conformed to developmental patterns (Fig. 5, 6, 7, 8).

Whereas small embryoid bodies usually contained embryonal carcinoma, mesenchyme and endoderm, those over 6 mm in diameter usually lacked embryonal carcinoma. It had either completely differentiated into ectodermal structures or it had succumbed. presumably because it was far removed from nutritional sources. These large embryoid bodies, which had been derived from embryonal carcinoma, were transplanted singly subcutaneously into animals to determine their effect upon the host (Pierce *et al.*, 1960). If these well differentiated appearing tissues were benign, they should occasion their host no difficulty, but if they were malignant, they should grow progressively and eventually destroy the animal. These cysts gradually underwent a conversion to what appeared to be the equivalent of benign dermoid cysts of the ovary. They remained in subcutaneous space for about 25% of the life-span of the animal and caused it no apparent difficulty. The cells and tissues were beautifully organized and functional. It was concluded that the end point of the differentiation of embryonal carcinoma cells was well differentiated postmitotic functional cells incapable of forming tumors upon transplantation. In other words, they were benign (Pierce *et al.*, 1960).

Squamous Cell Carcinomas

Since the observations described above were made in germinal tumors, and it might be argued that germinal tumors are a class unto themselves, it was decided to see if other tumors behaved similarly.

For the next series of experiments, a squamous cell carcinoma of the lip of an Irish rat, isolated by Dr. K. Snell, was studied (Pierce and Wallace, 1971). This tumor was slowly, yet progressively growing and was composed of two cell types. The first was an exceedingly undifferentiated cell that grew in large masses which surrounded smaller aggregates of squamous cells. The latter, long known as squamous pearls to pathologists, closely resemble normal squamous cells. If a small fragment of normal squamous epithelium is introduced beneath the skin of an animal, it will form a cystic structure composed of a basal layer with all levels of squamous differentiation leading to a necrotic center filled with keratotic material.

Animals bearing this squamous cell carcinoma were given pulses of tritium-labeled thymidine and examined by autoradiography with light and electron microscopes at prescribed times to determine which cells synthesized DNA and to follow their subsequent fate. It turned out that two hours after an injection of tritium-labeled thymidine, in a survey of 5,410 labeled undifferentiated cells, 779 pearls were encountered and only 5 labeled cells

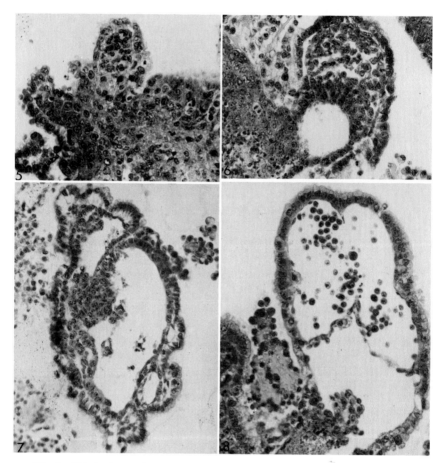

Fig. 5. *This is the surface of an explant grown on sponge matrix with two papillary projections composed of visceral yolk sac and embryonal carcinoma. Although we have traced the development of complex embryoid bodies from this type of structure, it is impossible to say that this one would have developed into a free floating cystic embryoid body. (X 260)*

Fig. 6. *This is a later stage in the genesis of an embryoid body in vitro than that illustrated in Fig. 5. The structure is larger, projects farther from the surface of the explant and contains mesenchyme that lies between the investing yolk sac and embryonal carcinoma from which it is differentiating. (X 260)*

Fig. 7. *This is an embryoid body which differentiated in sponge matrix cultures and in addition to yolk sac, embryonal carcinoma and mesenchyme has sinusoidal spaces beneath the yolk sac. (X 260)*

Fig. 8. *This embryoid body differentiating in sponge matrix cultures after 90 days in vitro differs from that in Fig. 7 in that it has immature nucleated hematopoietic cells in the capillary sinuses. (X 260)*

were observed in them. Contrast this to the situation at 96 hours: 219 labeled squamous cells were found in 733 pearls. These data were supportive of those of Frankfurt (1967), and it was concluded that the growth of this tumor was dependent upon the proliferation of undifferentiated cells which lay outside of the pearl, and that growth of the pearl was dependent upon the migration of undifferentiated cells into the pearl with subsequent differentiation. To be sure of the latter, the experiments were repeated and autoradiography with the electron microscope was done to ensure that the labeled cells inside of the pearl had differentiated. Whereas the labeled cells outside of the pearl lacked the features of squamous differentiation, labeled cells that had migrated into the pearl were characterized by the ultrastructural features of well differentiated squamous cells. Thus, it was concluded that undifferentiated stem cells had migrated into the pearls and had in fact differentiated. In order to determine whether or not they had differentiated to a benign form, 78 pearls were dissected from their tumors using a dissecting microscope and transplanted into compatible hosts. Grafts of equal amounts of undifferentiated tumors were treated similarly. Whereas none of the pearls developed into a tumor, about a third of the transplants of the undifferentiated cells formed squamous cell carcinomas (Pierce and Wallace, 1971).

It was concluded from these experiments that whereas the teratocarcinoma was a caricature of embryogenesis, the squamous cell carcinoma was a caricature of tissue genesis.

Chondrosarcoma

Similar experiments were performed upon a chondrosarcoma of the hamster (Pierce et al., 1973). These experiments have not been as satisfactory as those with the teratocarcinoma and squamous cell carcinoma because it has been impossible to isolate aggregates of differentiated tissues to test their neoplastic potentials. But the lessons learned are similar. Undifferentiated cells that synthesize DNA are associated with little chondromucoprotein and those that do not synthesize DNA are separated by large amounts of it (Fig. 9, 10).

When the chondrosarcoma was examined with the electron microscope, unlike the situation in embryonal carcinoma or in the squamous cell carcinomas where the stem cells were exceedingly undifferentiated, these stem cells were beautifully differentiated with a complex Golgi apparatus and well differentiated cisternal profiles of rough endoplasmic reticulum (Fig. 11). Had these cells been observed in the teratocarcinoma, they would have been considered well differentiated cells and their stem cell capabilities would have been missed. Differentiation of these stem cells was recognized by production of large amounts of chondromucoprotein. This was, then, further confirmation that tumors were caricatures of tissue genesis.

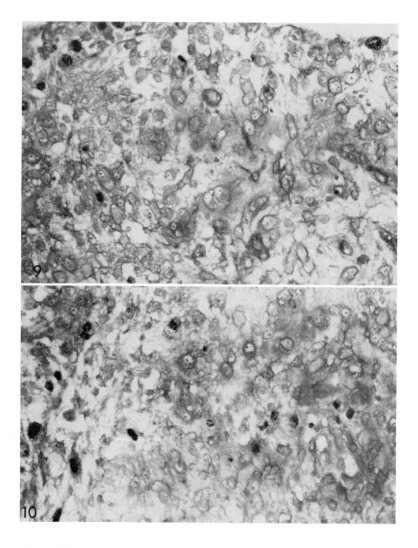

Fig. 9. *This is an autoradiogram of the chondrosarcoma taken two hours after a pulse of tritium labeled thymidine. The cells in the center of the field are not synthesizing DNA and are widely separated by large amounts of chondromucoprotein. Numerous labeled cells are situated in the close packed cells with little matrix. (X 400)*

Fig. 10. *This is an autoradiogram of the chondrosarcoma taken 96 hours after a pulse of tritium labeled thymidine. Labeled cells are found in close packed cellular areas peripheral to the cells associated with large amounts of matrix. There are labeled cells in the latter areas suggesting that cells migrate into these areas and differentiate and function. (X 400)*

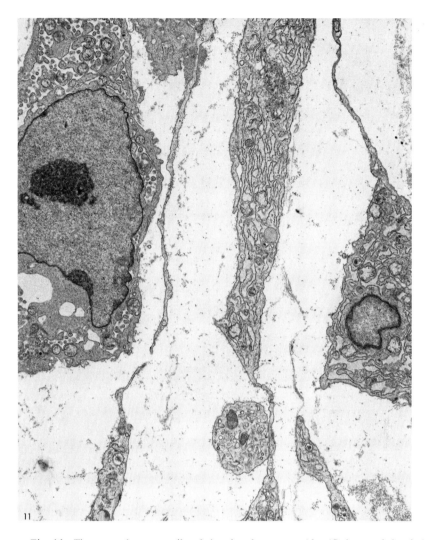

Fig. 11. *These are the stem cells of the chondrosarcoma identified as such by their ability to synthesize DNA. Note the extreme development of irregular and cisternal rough endoplasmic reticulum. These cells are cytologically indistinguishable from those not synthesizing DNA. The latter synthesize large amounts of chondromucoprotein. (X 12,600)*

Breast Carcinoma

To obviate the argument that these studies had all been performed upon transplantable tumors which might have been highly selected and no longer representative of the state of affairs in primary tumors, it was decided to study the cellular kinetics of primary adenocarcinomas of the breast of mice (Wylie *et al.*, 1973). In addition, we wished to determine whether or not a tumor induced by a virus would follow the same rules of differentiation as observed in tumors of presumably non-viral origin.

As background to this problem, it should be noted that Mendelsohn (1962) observed a large number of nonproliferating cells in mammary tumors of mice that had been perfused continuously for five days with tritium-labeled thymidine. He postulated that these nonproliferating cells were probably stem cells arrested in G-0. The experiments of Mendelsohn were repeated, animals with spontaneous adenocarcinomas of the breast were perfused with tritium-labeled thymidine for five days and the tumors were examined by autoradiography with the electron microscope (Wylie *et al.*, 1973). A wide range of differentiation was present ranging from cells lining acini with secretion droplets and well differentiated organelles to poorly differentiated stem cells (Fig. 12). Attention was paid to the degree of differentiation of unlabeled cells that had not synthesized DNA for the five days. As a positive control the degrees of differentiation of cells capable of synthesizing DNA was determined by pulsing other tumors with tritium-labeled thymidine and describing the ultrastructural characteristics of the labeled cells. The results of the experiment are summarized in Table I and indicate in confirmation of the observations of Mendelsohn that a significant number of the nonproliferating cells are stem cells which apparently do not synthesize DNA. On the other hand, there are a sizable number of cells which are not synthesizing DNA and which are exceedingly well differentiated, a degree of differentiation found to be incompatible with synthesis of DNA in the pulse experiments (Fig. 13). It was concluded that they were postmitotic as the result of differentiation.

Cells actively synthesizing virus usually lacked secretion droplets and organotypic differentiation. This was not to imply that they were undifferentiated, rather viral production appeared to have its own unique appearance as though synthesis of an RNA virus was a differentiated function in its own right (Wylie *et al.*, 1973). Thus, it was concluded that the processes of differentiation in a tumor induced by an RNA virus can control the expression of the viral genome but whether or not the viral genome is still present in the benign appearing cells has not been determined. We have attempted to answer this question in the teratoma system by infecting the multipotential embryonal carcinoma cells with SV_{40} virus and determining whether or not the expression of this DNA viral genome can be controlled by the processes of differentiation and whether or not we can obtain populations of benign cells containing a repressed viral genome (Lehman *et al.*, 1973).

Discussion

It is now clear, particularly from the work of J. and E. Miller (1971), that the tissue, organ and species specificity of certain chemical carcinogens is

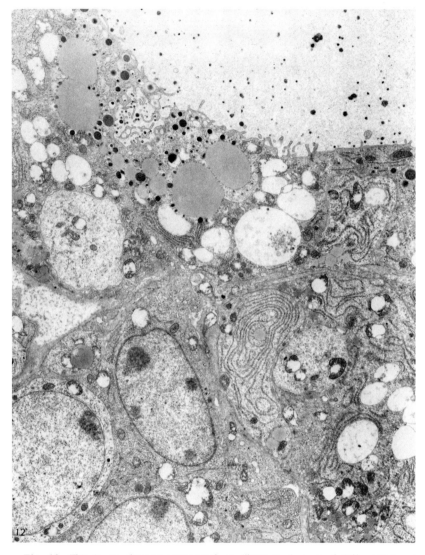

Fig. 12. *This is an electron micrograph to illustrate degrees of differentiation in mammary carcinoma cells. The cells lining the acinus have terminal bars, microvilli, secretion granules of two types, well developed Golgi complexes and rough endoplasmic reticulum. Contrast this appearance to that of cells in the lower left hand corner that have scant cytoplasm and few membranous organelles. They are undifferentiated stem cells. (X 11,800)*

due to the fact that the agents must be metabolized to an active carcinogenic form. If a species lacks the appropriate enzymes, conversion will not occur and the species will be resistant to carcinogenesis. If it has the enzymes, then

TABLE I
Degree of Differentiation

Tumor	+	++	+++
	% of Cells	% of Cells	% of Cells
A	55 (51)	41 (38)	12 (11)
B	59 (36)	84 (51)	23 (14)
C	15 (24)	26 (41)	22 (35)

depending upon the route of excretion of the activated chemicals and other factors, exposed tissues and organs will become neoplastic. The problem is to determine which cells are the target of carcinogenesis. From studies of carcinogenesis *in vitro* using chemicals and viruses, it is clear that fibroblasts give rise to fibrosarcomas. Carcinogenesis of the skin gives rise to skin tumors and the primordial germ cell, when it becomes neoplastic, gives rise to a teratocarcinoma. Thus, it can be deduced that the target cell in carcinogenesis is a stem cell in its own right. In other words, it is a mitotically active cell determined for a particular differentiation. Since after carcinogenesis it produces a caricature of the normal tissue, one could suppose that the essential change was an alteration of control mechanisms. The data do not rule out the possibility that a mutation in a control operon would be responsible for the change, but since so much evidence supports the notion that we are dealing with the process of differentiation, I would propose that there has been a change of control of genome converting the normal stem cell to a malignant stem cell.

If this is true, then it logically follows that the normal genome must contain all of the information necessary for the expression of the malignant phenotype (Pierce and Johnson, 1971).

The malignant phenotype is difficult to define or describe. As diagnosticians we utilize a combination of factors to arrive at a diagnosis of malignancy, among the features that would have to be present in the normal genome or at least be part of normal cellular function would be the attributes of proliferation at the expense of differentiation, usurpation of the logistics of the host, invasion and metastasis. These are the attributes that lead to the demise of the host, and it is interesting that each is expressed at some time in embryonic life. The fertilized egg proliferates without much overt evidence of differentiation, all fetuses have preferential treatment at the expense of the maternal organism. Invasion connotes aggression which relates to cancer, the disease. In developmental biology it is the equivalent of migration. Neural crest cells invade or migrate into the surrounding mesenchyme and are distributed widely throughout the body. The germ cells of the bird develop in the extraembryonic endoderm and metastasize via the blood stream to the gonads. The reticuloendothelial system metastasizes to the lymph nodes, spleen, etc. Thus, it would appear that the attributes by which we recognize

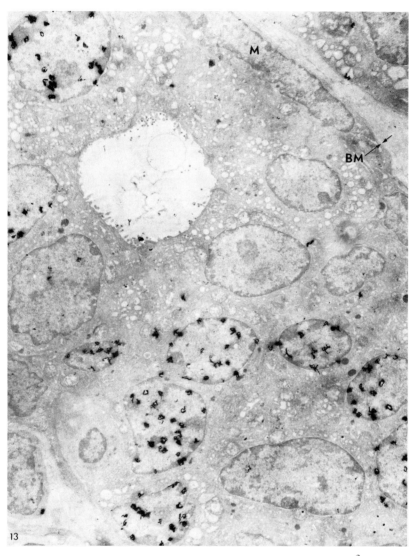

Fig. 13. *This electron autoradiogram (from an animal perfused with ³H-thymidine for five days), taken at low magnification, illustrates an acinus and solid tumor separated from stroma by a basement membrane (BM). Note the myoepithelial cell (M). Silver grains overlie the nuclei of cells synthesizing DNA. Unlabeled cells have not synthesized DNA for five days. Note the variations in degrees of differentiation of the cells as evidenced by the presence of secretion granules, degrees of development of rough endoplasmic reticulum, Golgi complexes, and amount of polysomes and free ribosomes. (X 7,800)*

malignancy are expressed at some time in normal development, but the summation of these effects operating as they are at levels of control incompatible with the needs of the host leads to such disproportionate growth that the host cannot support it. The result is cachexia and death. This is one of the essential features of malignancy.

If we are to make important inputs into directing differentiation of malignant tumor cells, understanding of differentiation as a process will be required as well as information leading to the manipulation of this process.

Certain key pieces of information form a base line for these considerations. The nuclear transplant experiments, those of King and McKinnell (1960) and King and DiBerardino (1965), demonstrate for one malignant system that cytoplasm has the capacity to control an integrated DNA viral oncogenic genome. In these experiments the nucleus of an adenocarcinoma cell of the frog, which is induced by a herpes 2-like virus, was transplanted into the activated but enucleated cytoplasm of a frog egg. Rather than disorganized growth until the yolk of the egg had been consumed, development proceeded along relatively normal lines. It remains to be shown whether or not the viral genome is still present in the embryonic cells. Although development was not complete as Braun (1972) has shown in cloning teratoma cells of plants, the conclusion is justified that cytoplasm controls genome.

Our confirmation of Mendelsohn's postulate that there are stem cells arrested in G-0 in tumors is an indication that a malignant cell's environment can control its behavior, a fact recognized in clinical practice as dormancy. A 48 year old woman may present with an adenocarcinoma of the breast which is apparently cured by radical surgery. For 15 years she may be well and then a small pea-sized nodule develops in the mastectomy scar and in spite of all treatment, she succumbs in a relatively short period of time to a tumor identical with the original tumor. Obviously a few tumor cells were trapped in an environment which precluded the malignant cells from expressing their phenotype for 15 years, or until a critical mass developed. We need to know about the intimate environment of cells.

Test systems need to be developed if we are to assay for factors and for environmental circumstances which will control neoplastic growth. Braun (1972) has presented evidence that in plant systems neoplastic cells have acquired the ability to produce two growth promoting substances characteristic of rapidly dividing embryonic plant cells. Search must be made for agents which autostimulate growth of mammalian cells. Naturally occurring agents must be searched for in the cellular environment and for which hormones and vitamins are model systems.

We have developed cell lines of the multipotential cells of teratocarcinomas to test inductive effects (Speers et al., 1973). This type of experiment is difficult of interpretation. For one thing, efforts in vitro are usually

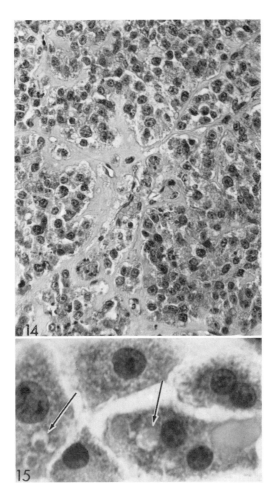

Fig. 14. *This is a subcutaneous transplant of parietal yolk sac carcinoma cells illustrating the massive amounts of basement membrane material synthesized by the tumor. (X 280)*

Fig. 15. *These are part of a monolayer of cells from the tumor illustrated in Fig. 13. Note the balls of basement membrane in the cells. (X 420)*

directed at making things grow. All kinds of nonspecific effects can preclude growth. We have used the multipotential system to try and direct the differentiation of embryonal carcinoma from one malignant cell type to another which will be able to grow competitively and therefore be observed. The initial experiments were done *in vivo*, were particularly laborious and gloriously unsuccessful. We hope that we will have better luck with the *in vitro* test systems recently developed (Speers *et al.*, 1973).

There are aspects of neoplasia which are effective as models of normal differentiating systems. In this regard, the myelomas have been particularly useful in obtaining monoclonal lines of immunoglobulin for molecular analysis. Similarly we have used a parietal yolk sac carcinoma, one of the malignant lines of cells that differentiate from embryonal carcinoma, to study the synthesis of basement membrane (Fig. 14, 15). Parietal yolk sac epithelium synthesizes Reichert's basement membrane of the mouse embryo (Pierce *et al.*, 1962, 1963; Pierce, 1970; Pierce and Johnson, 1971). We have been able to obtain the basement membrane molecule free of contaminating collagen, study its molecular configurations, and have observed that epithelial cells in response to injury synthesize large amounts of this material, presumably as a productive mechanism. Judicious use of tumors as easily manipulated models of normal situations is therefore indicated for studying normal development.

The identification of benign cells derived from malignant stem cells has led to a developmental concept of cancer. This concept supposes that normal stem cells are the target in carcinogenesis. As a result of alteration of genomic controls with repression and derepression of genomic loci, and providing that the cellular milieu is appropriate, the malignant phenotype eventuates. The neoplasm is a caricature of normal tissue, and a caricature of the process of cell renewal. The model systems of differentiation that developmental biologists have been waiting for have arrived packaged under the names, carcinogenesis and transformation.

REFERENCES

1. Boveri, T. (1912). Zur Frage der entwicklung maligner tumoren. Jena. (1929). (English Translation). *The Origin of Malignant Tumors,* p. 119, Williams and Wilkins, Baltimore.

2. Braun, A. C. (1972). The usefulness of plant tumor systems for studying the basic cellular mechanisms that underlie neoplastic growth generally. In: *Cell Differentiation,* pp. 115-119, Harris, R., Allin, P. and Viza, D., eds., Munksgaard.

3. Dixon, F. J., Jr., and Moore, R. A. (1953). Testicular tumors – A clinicopathologic study. *Cancer 6*, 427-454.

4. Frankfurt, O. S. (1967). Mitotic cycle and cell differentiation in squamous cell carcinomas. *Intern. J. Cancer 2*, 304-310.

5. King, T. J., and McKinnell, R. G. (1960). In: *Cell Physiology of Neoplasia*, p. 691, University of Texas Press, Austin, Texas.

6. King, T. J., and DiBerardino, M. A. (1965). Transplantation of nuclei from the frog renal adenocarcinoma. I. Development of tumor nuclear-transplant embryos. *Ann. N. Y. Acad. Sci. 126*, 115-126.

7. Kleinsmith, L. J., and Pierce, G. B. (1964). Multipotentiality of single embryonal carcinoma cells. *Cancer Res. 24*, 1544-1551.

8. Lehman, J. M., Swartzendruber, D., and Pierce, G. B. (1973). Unpublished.

9. Markert, C. L. (1968). Neoplasia: A disease of cell differentiation. *Cancer Res. 28*, 1908-1914.

10. Melicow, M. M. (1940). Embryoma of testis. Report of a case and a classification of neoplasms of the testis. *J. Virol. 44*, 333-344.

11. Mendelsohn, M. L. (1962). Autoradiographic analysis of cell proliferation in spontaneous breast cancer of C_3H mouse. III. The growth fraction. *J. Nat. Cancer Inst. 28*, 1015-1029.

12. Miller, J. A. (1970). Carcinogenesis by chemicals – An overview. G.H.A. Clewes Memorial Lecture. *Cancer Res. 30*, 559-576.

13. Peyron, A. (1939). Faits nouveau relatifs a l'vrigine et a' l' histogenese dis embryomes. *Bull. du Cancer 28*, 658-681.

14. Pierce, G. B. (1967). Teratocarcinoma: Model for a developmental concept of cancer. In: *Current Topics in Developmental Biology, Vol. 2*, pp. 223-246, Moscona, A. A., and Monroy, A., eds., Academic Press, Inc., New York.

15. Pierce, G. B. (1970). Epithelial basement membrane: Origin, development and role in disease. In: *Chemistry and Molecular Biolory of the Intercellular Matrix, Vol. 1*, pp. 471-506, Balazs, ed., Academic Press, New York.

16. Pierce, G. B., and Beals, T. F. (1964). The ultrastructure of primordial germinal cells of the fetal testes and of embryonal carcinoma cells of mice. *Cancer Res. 24*, 1553-1567.

17. Pierce, G. B., and Dixon, F. J. (1959). Testicular teratomas. I. The demonstration of teratogenesis by metamorphosis of multipotential cells. *Cancer 12*, 573.

18. Pierce, G. B., and Johnson, L. D. (1971). Differentiation and cancer. *In Vitro 7*, 140-145.

19. Pierce, G. B., and Midgley, A. R. (1963). The origin and function of human syncytiotrophoblastic giant cells. *Amer. J. Path. 43*, 153-173.

20. Pierce, G. B., and Verney, E. L. (1961). An *in vitro* and *in vivo* study of differentiation in teratocarcinomas. *Cancer 14*, 1017-1029.

21. Pierce, G. B. and Wallace, C. (1971). Differentiation of malignant to benign cells. *Cancer Res. 31*, 127-134.

22. Pierce, G. B., Dixon, F. J., and Verney, E. L. (1960). Teratocarcinogenic and tissue forming potentials of the cell types comprising neoplastic embryoid bodies. *Lab Invest. 9*, 583-602.

23. Pierce, G. B., Dixon, F. J., and Verney, E. L. (1959). Endocrine function of a heterotransplantable human embryonal carcinoma. *Arch. Path. 67*, 204.

24. Pierce, G. B., Dixon, F. J., and Verney, E. L. (1958). The biology of testicular cancer. II. Endocrinology of transplanted tumors. *Cancer Res. 18*, 204-206.

25. Pierce, G. B., Huffer, W. E., and Wylie, C. V. (1973). Unpublished.

26. Pierce, G. B., Midgley, A. R., and Sri Ram, J. (1963). The histogenesis of basement membranes. *J. Exp. Med. 117*, 339-348.

27. Pierce, G. B., Midgley, A. R., Feldman, J. D., and Sri Ram, J. (1962). Parietal yolk sac carcinoma. Clue to the histogenesis of Reichert's membrane of the mouse embryo. *Amer. J. Path. 41*, 549-566.

28. Pierce, G. B., Stevens, L. C., and Nakane, P. K. (1967). Ultrastructural analysis of the early development of teratocarcinomas. *J. Nat. Cancer Inst. 39*, 755-773.

29. Rowe, W. P. (1973). Genetic factors in the natural history of murine leukemia virus infection. *Cancer Res., 33*, 3061-3068.

30. Speers, W. C., Lehman, J. M., and Pierce, G. B. (1973). Spontaneous differentiation of mouse teratocarcinoma in tissue culture. *Proc. Am. Assoc. Cancer Res. 14*, 112.

31. Stevens, L. C. (1960). Embryonic potency of embryoid bodies derived from a transplantable testicular teratoma of the mouse. *Dev. Biol. 2*, 285-297.

32. Stevens, L. C. (1962). The biology of teratomas including evidence indicating their origin from primordial germ cells. *Ann. Biol. 1*, (11-12), 41.

33. Stevens, L. C. (1964). Experimental production of testicular teratomas in mice. *Proc. Nat. Acad. Sci. US 40,* 1081.

34. Stevens, L. C. (1967). Origin of testicular teratomas from primordial germ cells in mice. *J. Nat. Cancer Inst. 38,* 549-552.

35. Stevens, L. C., and Little, C. C. (1954). Spontaneous testicular teratomas in inbred strain of mice. *Proc. Nat. Acad. Sci. US 40,* 1080-1087.

36. Todaro, G. J., and Huebner, R. J. (1972). The viral oncogene hypothesis: New evidence. *Proc. Nat. Acad. Sci. US 69,* 1009-1015.

37. Verney, E. L., and Pierce, G. B. (1959). The biology of testicular tumors. III. Heterotransplanted choriocarcinoma. *Cancer Res. 19,* 633.

38. Wylie, C. V., Nakane, P. K., and Pierce, G. B. (1973). Degrees of differentiation in nonproliferating cells of mammary carcinoma. *Differentiation 1,* 11-20.

ACKNOWLEDGEMENTS

The author wishes to express his appreciation to Mr. Alan Jones for his excellent technical assistance. Figures 1, 2, 3, 5, 6 and 7 are reproduced from Pierce and Verney, *Cancer, 14,* 1017, 1961 and Figures 12 and 13 from Wylie *et al., Differentiation, 1,* 11.

MECHANISMS UNDERLYING TUMOR TRANSFORMATION AND TUMOR REVERSAL IN CROWN-GALL, A NEOPLASTIC DISEASE AT HIGHER PLANTS

Frederick Meins, Jr.

Department of Biology
Princeton University
Princeton, New Jersey 08540

Tumor cells have a pronounced capacity for autonomous growth which persists when the cell divides. Crown-gall, a neoplastic disease of higher plants, provides a model system particularly suited to studies of both tumor autonomy and inheritance of the tumor state. The key advantages of this system are that, unlike tumors of animal origin, the physiological basis for autonomy of crown-gall tumors has been established, and that genetic and epigenetic mechanisms for transformation may be distinguished by regenerating entire plants from progeny of individual plant cells (Braun, 1961). Another advantage for biochemical studies is that heritable changes in tumor autonomy and development can be induced at will in culture using chemically defined substances (Meins, 1971).

In this paper, I will review the salient properties of the crown-gall system and then describe experiments supporting the concept that tumor transformation in plants has as its basis persistent epigenetic changes of the type encountered in normal development. The reader is referred to a recent review volume for a detailed treatment of plant tumor diseases (Braun, 1972a).

Tumor Induction and Stability of the Tumor Phenotype

Crown-gall tumors result when plant cells conditioned by wounding are exposed to a tumor-inducing principle (TIP) which is produced by the crown-gall bacterium, *Agrobacterium tumefaciens* (Braun, 1947; Braun and

Mandle, 1948). Once transformation is accomplished, crown-gall tumor cells continue to proliferate autonomously in the absence of the inciting bacteria (Braun and White, 1943) indicating that treatment with the bacterium induces a true neoplastic transformation.

The type of tumor obtained depends upon the strain of crown-gall bacterium and the species of host plant used. For example, in tobacco, the B6 strain of the bacterium induces rapidly growing tumors which show little capacity for differentiation. In contrast, the moderately virulent T37 strain induces complex teratomas consisting of a chaotic arrangement of highly abnormal leaves and stems (Braun, 1953). Sterile explants removed from either organized teratomas or unorganized tumors retain their characteristic pattern of development and capacity for autonomous growth in a host as well as in axenic culture indicating that the distinctive tumor phenotypes are inherited characters. Moreover, cloned teratoma tissues are indistinguishable from the teratoma tissue from which they were derived. Thus, the capacity for teratomatous development is a heritable property of individual multi-potential teratoma cells (Braun, 1959).

The teratoma phenotype is remarkably stable; tissues that have been propagated continuously for over 25 years in culture still exhibit their characteristic pattern of abnormal development. Nonetheless, these tissues occasionally give rise to highly stable variants which have been subcultured for 14 years on the same growth medium as the tissue from which they were derived without reverting to the parental phenotype. The two variants studied in detail differ from the parent line and each other in their pattern of development and in their nutritional requirements (Meins, 1969, 1971). One variant, *snowy*, is glistening white in appearance and consists of two distinct tissue layers, a cortex composed of hyphae-like projections of filamentous cells, and a core composed of parenchyma cells. The other variant, *unorganized*, is composed of a mass of undifferentiated cells and is morphologically indistinguishable from fully transformed tumor tissue obtained by inoculating tobacco with the B6 strain of the crown-gall bacterium. Comparative studies of the variants and several cloned lines of organized tumor tissues indicate that crown-gall tumors, like animal tumors, can undergo heritable variation to yield diverse tumor types with different capacities for biosynthesis and differentiation.

The Physiological Basis for Tumor Autonomy

The enlargement and division of higher plant cells requires the concerted action of two types of growth regulators: an auxin and a cell division factor. Thus, tobacco pith parenchyma tissues require an exogeneous source of auxin and a cell division factor such as a 6-substituted purine cytokinin or a cytokinesin for continuous growth in culture (Jablonski and Skoog, 1954;

Miller *et al.*, 1955; Wood *et al.*, 1972). In striking contrast, crown-gall tissues in culture can grow in the absence of these factors which, as a result of transformation, they are now able to produce (Braun, 1958).

There is a close parallel between the nutritional requirements of crown-gall tissues in culture and their capacity for autonomous growth. In a graded series of *Vinca rosea* tumor tissues with increasing growth rate, there is a progressive decrease in their requirement in culture for several growth factors such as glutamine, asparagine, purines, pyrimidines, *myo*-inositol, auxin and a cell division factor. These results have led Braun (1958) to conclude that the autonomous growth of crown-gall cells has as its basis the persistent activation of several diverse biosynthetic capacities required for the proliferation and enlargement of plant cells. These biosynthetic capacities, usually blocked in normal cells from which the tumor cells are derived, are progressively activated with increasing capacity for autonomous growth.

There is considerable evidence that activation of these capacities is sufficient to account for the autonomous growth of the tumor cell. Normal tobacco tissues in culture occasionally lose their requirement for auxin and a cell division factor (Fox, 1963) and, in this regard, are indistinguishable from crown-gall tumor cells. When these tissues are grafted onto tobacco plants they form typical non-selflimiting tumors that are serially transplantable (Limasset and Gautheret, 1950). This close correlation between growth factor production and tumor autonomy is also encountered in genetic hybrid tumors of tobacco.

Interspecific hybrids of certain *Nicotiana* species readily form spontaneous tumors. This capacity for autonomous growth results from particular combinations of genes contributed by the parental species and is not due to a transmissable causal agent (Smith, 1972). Schaeffer and Smith (1963) have compared the nutritional requirements in culture of tissues excised from two non-tumorous species, *N. suaveolens* and *N. langsdorffii* with tissues excised from a tumorous *N. suaveolens x langsdorffii* hybrid. They found that tissues from the parental species were unable to grow in the absence of added auxin and a cell division factor while the tumorous hybrid tissues, like crown-gall transformed tissues, could grow without an exogenous supply of these growth factors. The fact that the capacity for autonomous growth, regardless of its proximal cause, is associated with production of auxin and a cell division factor is strong evidence that tumor autonomy results from the activation of normally repressed biosynthetic capacities of the plant cell.

Reversibility of the Tumor State

The fundamental question is how production of growth factors, once initiated, is inherited by individual tumor cells. A fruitful approach to this question has come from the demonstration that the crown-gall transformation is potentially reversible. Teratoma cells of tobacco retain the capacity to form some structures found in the normal plant suggesting that these cells might lose their tumor properties altogether when subjected to the proper inductive environment. To accomplish this, Braun (1959) grafted cloned teratoma tissues onto the cut stem tips of tobacco plants from which auxillary buds had been removed. .The grafts developed typical teratomas with numerous abnormal shoots. Tips from abnormal shoots grafted a second time grew more rapidly and became more normal in appearance. When the tips of these shoots were grafted a third time, normal shoots developed which eventually flowered and set seed. These seeds germinated and developed into complete, normal tobacco plants.

To eliminate the possibility that the normal shoots developed from the host plant, Braun (in press) regenerated shoots from cloned teratoma tissues of Havana 38 origin which he then grafted onto a Turkish tobacco host. As in the earlier experiment, some shoots eventually flowered and set seed. In every case, the complete plants obtained were of the Havana type and hence derived from the teratoma tissue (Fig. 1).

These experiments demonstrate that autonomous teratoma cells can lose their tumor properties in a gradual process and develop once again in a normal fashion. Therefore, the transformation process in crown-gall is potentially reversible and not due to permanent changes in the host cell genome. Moreover, the teratoma cell is totipotent since it retains all the information necessary to regenerate the entire plant.

Similar conclusions can be drawn from regeneration experiments using fully transformed tumor cells of tobacco that do not normally organize in culture (Sacristán and Melchers, 1969). Highly aneuploid, abnormal plants were regenerated from these tumor tissues in culture. Sterile tissue fragments obtained from these plants were no longer able to grow in culture without added auxin and cytokinin nor were the plants any more susceptible to tumor transformation by crown-gall bacteria or wounding than seed grown plants. These results suggest that the fully transformed tumor cell, although capable of non-tumorous growth, is no longer totipotent. However, it is unlikely that genetic changes are involved in the transformation and recovery processes since normal tissues subjected to the same treatment to induce regeneration also give rise to abnormal and highly aneuploid plants (Sacristán and Melchers, 1969). The important point is that these experiments, like those of Braun, indicate that the tumor state is potentially reversible and that tumor reversal occurs with great regularity.

An Oncogenic Principle in Teratoma Cells

Two mechanisms could, in principle, account for both the heritability and potential reversibility of the tumor state (Braun, 1969; Meins, 1972): *gene addition*, in which transformation results from infection of the host cell with

Fig. 1. *Tumor recovery using genetically marked teratoma tissues of tobacco. A. Shoots regenerated from cloned teratoma tissue of Havana 38 origin. B. Havana shoots grafted onto a Turkish tobacco host. C. Later stage in the development of the graft. D. Shoot grafted a second time on the Turkish host which eventually flowered and set fertile seed. Note differences in leaf morphology of graft and host. Shoots exhibit the Havana phenotype characteristic of the teratoma tissue. (Braun, in preparation).*

a self-replicating entity such as RNA, DNA, or a virus particle; and, *epigenetic modification*, in which the tumor state is perpetuated by heritable changes in the expression of host-cell genes. Although there is, at present, no clear cut evidence favoring one mechanism over the other (Beardsley 1972), the gene addition hypothesis is supported indirectly by the finding that tumor cells may contain base sequences present in *A. tumefaciens* DNA (Schilperoort *et al.*, 1967; Quétier *et al.*, 1969) as well as several bacterial gene products (Milo and Srivastava, 1969; Chadha and Srivastava, 1971). If transformation does result from the persistent replication of a foreign entity, then crown-gall tumor cells should contain an *oncogenic principle* either identical to, or derived from the agent which initiated the transformation. Two types of experiments suggest this is the case. First, tumorous overgrowths, of uncertain origin, sometimes arise at the graft union of normal and tumor tissues (deRopp, 1947; Camus and Gautheret, 1948). Second, X-irradiated tumor tissues of tobacco induce transplantable tumors in a tobacco host (Aaron-DaCunha, 1969).

To find out whether or not totipotent teratoma cells of tobacco contain an oncogenic principle, I grafted small explants of Turkish teratoma tissue near the vascular bundle in Havana 425 tobacco plants using different varieties of tobacco to distinguish between host and grafted tissues (Meins, 1973). Within two weeks after the graft was made, teratomatous overgrowths appeared on the host plant often at least 2 mm removed from the teratoma implant. Although the original implant usually did not grow, the newly formed overgrowths grew rapidly and were morphologically indistinguishable from teratomas induced by the T37 strain of the crown-gall bacterium (Fig. 2).

Induction of overgrowths required the presence of teratoma tissue. Overgrowths did not form when Turkish pith, Havana pith, or Havana vascular tissues were grafted onto the Havana host indicating that tumor induction is not due to interactions between the Havana and Turkish tissues or the grafting procedure itself. Since teratoma tissues produce auxin and cell division factors, it is possible that these substances trigger cell proliferation in the Havana host which thereafter is self-sustained. To test this hypothesis, lanolin containing 10mg/g. of an auxin, naphthalene acetic acid, and a cytokinin, kinetin, was implanted into Havana plants at the same site used in grafting. Although large unorganized overgrowths developed in response to this treatment, the overgrowths could not be serially propagated and hence were not autonomous. The induction of tumors thus appears to show a specific requirement for crown-gall tumor tissues.

The induced overgrowths were transplantable and autonomous. Small explants of overgrowth tissue grew into large tumors when grafted onto a Turkish host. These tumors have been serially transplanted three times over a period of four months and have continued to form non-self-limiting growths. Explants of the induced overgrowths could also be propagated in axenic culture on a basal medium that supports the growth of crown-gall tumor tissues but not normal Havana tissues. These findings indicate that the overgrowths, like crown-gall tumors, are true tumors capable of self-sustained growth.

Fig. 2. *Teratoma bearing the Havana 425 phenotype induced in a Havana host by a small implant of Turkish teratoma tissue. One month after grafting (Meins, 1973).*

Several lines of evidence show that induced tumors arise from normal cells and not from the teratoma implants used to induce the transformation. First, induced tumors were top grafted onto a Turkish host to induce development of normal shoots. In every case, the shoots exhibited the characteristic Havana phenotype. Second, histological sections of Havana host bearing teratoma grafts show a clear line of demarkation between implanted teratoma tissue and the host and verify that induced teratomas arise at a distance from the grafted tissue (Fig. 3). Finally, autonomous teratomas could also be induced in Havana stem segments in axenic culture where the teratoma explants were separated from the host by 0.22μ pore size Millipore filters. Since tobacco cells are far larger than the pore size of these filters, it is highly unlikely that tumors arising in the host are contaminated with cells derived from the implanted teratoma tissue.

The experiments presented in this section demonstrate that teratoma cells of tobacco contain an oncogenic principle in an active form that is readily transmitted to susceptible normal cells. This conclusion is consistent with the hypothesis that the tumor state results from expression of foreign genes replicating in synchrony with the host cell. However, there is at present no more reason to believe that this oncogenic principle is a self-replicating entity or that it is identical to TIP produced by crown-gall bacteria than that it is an

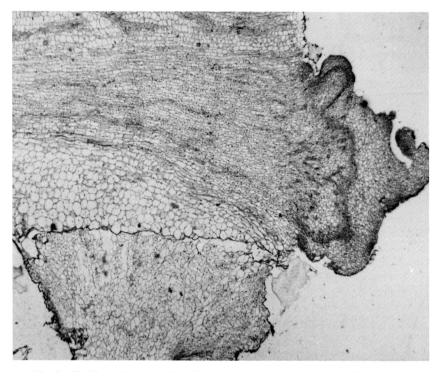

Fig. 3. *Histological section of a Havana host plant with implanted Turkish teratoma tissue. 16 days after grafting showing the distinct line demarking implanted tissue (right) and the developing teratoma derived from the host (left). (Meins, 1973).*

oncogenic factor produced by the teratoma cells as a result of transformation. The crucial experiment is to isolate and characterize this oncogenic principle. Since the principle can be transmitted through membrane filters and can be easily assayed in culture, this experiment is now feasible.

Heritable Activation of Growth Factor Production by Normal Cells

Gene addition and epigenetic modification are not mutually exclusive mechanisms for crown-gall transformation. For example, foreign genes introduced into the host cell during the inception phase of transformation could act either continuously or, alternatively, they could trigger persistent epigenetic modifications early in the transformation process which then maintain the tumor state. The essential point is that the epigenetic mechanism does not require the continued presence of foreign genes in the host cell and, moreover, is generally applicable whether or not transformation is due to a self-replicating entity. The epigenetic hypothesis makes specific predictions that can be tested experimentally, *viz.*, it should be possible to find cells that have become heritably but reversibly altered to produce those growth factors which are produced as a result of crown-gall transformation even in the absence of TIP or other infectious agents. I will present strong evidence that this is the case.

As pointed out earlier, pith parenchyma tissues of tobacco normally require an exogeneous supply of both auxin and a cell division factor, usually provided as the cytokinin, kinetin, for continued growth in culture. These tissues sometimes lose their requirement for either or both growth factors after a variable number of passages in culture. Thereafter the tissue can be propagated indefinitely without added growth factor. This type of heritable change, known as *habituation*, occurs in cultured tissues of a variety of plant species (Gautheret, 1955; Fox, 1963).

Habituated tissues produce significant amounts of the growth factor for which they are habituated (Gautheret, 1955; Fox, 1963; Wood *et al.*, 1969; Dyson and Hall, 1972; Einset and Skoog, 1973). The fact that intact plants also produce these growth factors suggests that habituation is due to the persistent activation of genes not normally expressed in culture. If this is the case, then habituation should be potentially reversible. Lutz (1971) was able to induce reversal of two cloned lines of tobacco habituated for both auxin and a cell division factor and reversal of auxin-habituation has been reported using tissues of *Crepis capillaris* (Sacristán and Wendt-Gallitelli, 1971) and tobacco (Sacristán and Melchers, 1969) that had not been cloned. These experiments, although suggestive, did not eliminate the possibility that the loss of the habituated character was due to cell selection or rare genetic mutations.

To get at this problem, Andy Binns, a graduate student in my laboratory, and I have carried out reversal experiments using a large number of cloned tobacco cells that have become cytokinin-habituated, i.e., able to grow continuously in culture without the cytokinin, kinetin (Binns and Meins, 1973). We isolated clones from a cytokinin-habituated tissue line which arose spontaneously from normal Havana 425 tobacco pith maintained in culture for two years. In our first experiment, 254 lines of single cell origin were isolated of which 38 lines were normal and 216 were cytokinin-habituated. The habituated clones have now been serially propagated for at least 54 cell doublings in the absence of kinetin indicating that cytokinin-habituation is inherited by individual cells.

Different habituated clones vary dramatically in their growth rate suggesting that the clones differ in degree of habituation. To eliminate the possibility that the clones simply differ in their intrinsic growth rate, we compared growth of each clone in the presence and absence of kinetin. The data were expressed as the ratio of growth without kinetin to growth with kinetin which has a value close to zero (range: 0 to 0.08) for normal tissues. A sample of 100 clones isolated from habituated tissue fell into three groups: 1) 8% were not habituated; 2) 37% grew indefinitely in the absence of kinetin but not as rapidly as in the presence of kinetin, and; 3) 55% grew rapidly in the absence of kinetin and were inhibited by added kinetin (Fig. 4). The inhibitory effect is of particular interest since it is commonly observed for tobacco tissues that produce sufficient amounts of cell division factors for rapid growth (Fox, 1963; Meins, 1969). Different cloned lines also varied over several orders of magnitude in the concentration of kinetin required for optimum growth (Fig. 5). Note that even the slowly growing clone 17H with a growth optimum close to that of normal tissues can grow indefinitely in the absence of kinetin. These results indicate that the different cloned lines differ strikingly in degree of cytokinin-habituation and suggest that the individual cells differ in this way as well. However to establish this point unequivocally the lines must be cloned a second time to eliminate the possibility that the cloned tissues consist of mixtures of habituated cells and normal cells arising by reversion.

Skoog and Miller (1957) have shown that complete tobacco plants could be obtained by treating cultured tissues with an appropriate mixture of auxin and kinetin. Using a modification of this technique we have, to date, obtained 62 complete tobacco plants from 19 different cytokinin-habituated lines of single cell origin. Forty-five of these have flowered and, so far, fertile seed has been obtained from 13 different clones (Binns and Meins, 1973).

The plants regenerated from both normal and habituated clones had thicker stems, larger leaves, and a darker pigmentation than seed-grown plants raised under comparable conditions. Morphological changes of this type are commonly observed in tobacco plants regenerated from cultured tissues and are correlated with polyploidization of the cells in culture (Murashige and Nakano, 1966; Sacristán, 1967). The important point is that although the cultured cells may have undergone genetic modification, fertile plants can be regenerated from the progeny of individual cytokinin-habituated cells. There-

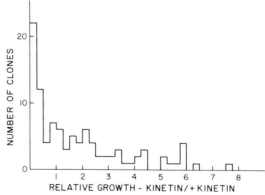

Fig. 4. *Relative growth rates of cytokinin-habituated clones in the presence and absence of the cytokinin, kinetin. Results shown for 100 independently isolated clones.*

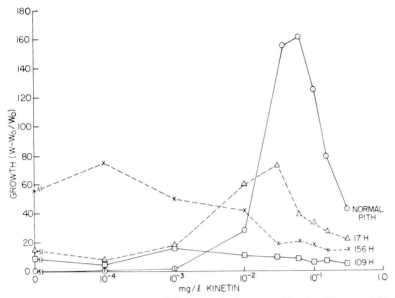

Fig. 5. *Dose response of several habituated clones to kinetin (Binns and Meins, 1973). Each point is the average of four replicates.*

fore, these cells retain all the information necessary to generate complete plants, i.e., the cells remain totipotent after the habituation process.

We tested pith explants from the regenerated plants for their ability to grow in culture in the absence of kinetin (Table 1). In every case, pith tissues derived from the habituated clones tested grew profusely in the presence of kinetin, but were no longer able to grow without kinetin and, in this regard, were indistinguishable from pith obtained from normal clones or seed-grown plants. Since tissues derived from cytokinin-habituated cells once again required cytokinin for growth in culture, we concluded that cytokinin-habituation is reversible and, therefore, not due to permanent genetic changes. The complete reversal experiment is summarized in Figure 6.

The experiments described demonstrate that the ability to produce cell division factors, a specific differentiated function of higher plant cells, is inherited by individual tobacco cells and is regularly reversible. Several lines of evidence support the hypothesis that this type of differentiation is the result of epigenetic modifications and not rare, random mutations: 1. Habituation involves activation of latent biosynthetic capacities of the normal cell. 2. The habituation process is regularly reversible and the cells remain totipotent. 3. Mutation of a small number of cells followed by selection is unlikely to explain our results since habituation occurs in tissues maintained on concentrations of kinetin that inhibit growth of habituated tissues but not normal tissues. 4. The fact that cloned lines differ in degree of habituation suggests

TABLE I.

Reversal of cytokinin-habituation in pith obtained from regenerated plants of clonal origin.

SOURCE OF PITH	GROWTH*		RELATIVE GROWTH
	A − Kinetin	B + Kinetin	A/B
Seed-grown Plants	20.±2.6(3)	1995±470(3)	0.010
Plants regenerated from normal clone 51N	4.7±1.5(3)	3311±821(2)	0.0014
Plants regenerated from cytokinin-habituated clones			
58H	10±2.2(5)	821±289(6)	0.012
66H	7.5±1.4(6)	468± 46(6)	0.016
82H	6.7±0.8(6)	1842±399(6)	0.0036
85H	15.±5.4(5)	2283±450(5)	0.0066

* mg. Fresh weight after 30 days. ± S.E. (N). Initial weight ca. 8 mg.

that, unlike most mutations, habituation is not an all or none change. 5. Finally, cytokinin-habituation occurs rapidly in culture. In preliminary experiments, we have estimated a habituation rate from measurements of changes in the frequency of habituated cells with time in culture. The lower limit for this rate is roughly 10^{-3} conversions/cell doubling. Although at present we cannot rule out genetic mutations that are directed and occur at high rates, such as paramutation in maize (Brink *et al.*, 1968) or anthocyanin variegation in *Nicotiana* hybrids (Smith and Sand, 1957), the best interpretation of our results appears to be that habituation is not the result of genetic changes.

CONCLUSIONS

The central principle underlying modern studies of development is that cell specialization results from orderly changes in gene expression and not from changes in the genes themselves (Davidson, 1968). This is supported by experiments showing that certain plant cells are totipotent (Vasil and Hildebrandt, 1965; Takebe *et al.*, 1971; Duffield *et al.*, 1972) and that nuclei from differentiated intestine cells of *Xenopus laevis* retain the necessary genes for development of the adult (Gurdon and Uehlinger, 1966). There are now several reports that cell determination, the commitment of a cell to specific development fates, can be inherited by individual chick and *Drosophila* cells (Cahn and Cahn, 1966; Coon, 1966; Gehring, 1968). Although the determined state is remarkably stable, in some cases cells may switch from one deter-

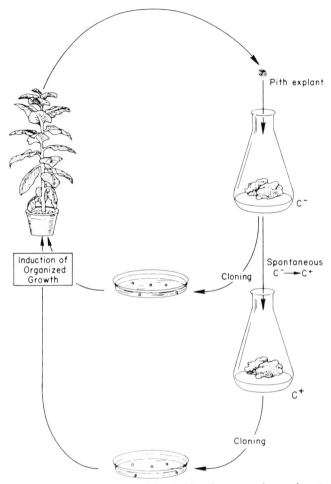

Fig. 6. Schematic representation of the habituation reversal experiment. C^-, cytokinin requiring tissues; C^+, cytokinin-habituated tissue able to produce cell division factors.

mined state to another (Gehring, 1968; Eguchi and Okada, 1973). Since cells that persistently express different determined states are thought to remain genetically equivalent, we must conclude that determination results from self-perpetuating epigenetic modifications.

Tumor transformation in crown-gall is a reversible process involving the heritable expression of specific latent capacities of the host cell. The autonomous tumor cells that result from this process are totipotent. Thus, there is striking similarity between tumor transformation and cell determination suggesting that both processes have a common epigenetic basis. This epigenetic theory, proposed most forcefully by Braun (1969, 1972b), maintains that tumor transformation in crown-gall, and perhaps generally, is essentially a process of anomolous differentiation which yields cells inappropriately locked into a pattern of neoplastic growth.

Studies of habituation provide strong support for this theory. Both cytokinin-habituation and transformation involve the heritable and potentially reversible conversion of a cytokinin-requiring cell into an autotrophic cell able to produce sufficient amounts of cell division factors for continuous growth in culture. The fact that habituation occurs in the absence of the crown-gall bacterium or added TIP indicates that persistence of an essential feature of the tumor state does not necessarily result from the expression of foreign genes introduced in the transformation process and carried thereafter by the host cell. Therefore, even in cases where transformation is initiated by a virus or a similar self-replicating entity, the primary event in transformation may, in fact, be the self-sustained expression of host genes.

The fundamental question is how self-sustained changes in the pattern of gene expression are maintained. A conceptual framework for dealing with this question is provided by steady state theories, first proposed by Delbrück (1949), and later applied to the problem of phenotypic stability in normal and neoplastic development. (Kacser, 1963; Monod and Jacob, 1961; Pitot and Heidelberger, 1963; Meins, 1972). Some systems of chemical reactions can exist in alternate self-perpetuating states. Thus, in principle, two genetically equivalent cells placed in the same environment, could inherit different phenotypes depending upon their past history. The stability and variety of phenotypes possible would depend upon the kinetic properties of the system and hence, ultimately, the genetic potentialities of the cell. The principle limitation in testing these theories has been the availability of experimental systems in which heritable, epigenetic changes can be regularly obtained and distinguished from genetic mutations and stable changes due to interactions between different cell types in a tissue.

Bednar and Linsmaier-Bednar (1971) have recently shown that substituted fluorenes, which are potent animal carcinogens, induce the cytokinin-habituation of normal tobacco cells in culture. Thus habituation, which can be induced and reversed under defined conditions in culture, and occurs in cells that can be tested directly for their developmental potentialities by regeneration of entire plants, provides a promising system for investigating the molecular basis for persistent epigenetic modifications encountered in cell determination and tumor transformation.

ACKNOWLEDGEMENTS

I gratefully acknowledge the technical assistance of Miss Mary Moore and the expert care of plants by Mr. Charles Feryok. This work was supported by grants from the Merck Company Foundation, The National Science Foundation, and the National Institutes of Health.

REFERENCES

1. Aaron-DaCunha, M. I. (1969). Sur la libération, par les rayons x, d'un principe tumorigène contenu dans les tissus de crown gall de tabac. *C.R. Acad. Sci. Paris D 268*, 318-321.

2. Beardsley, R. E. (1972). The inception phase in the crown-gall disease. *Progr. Exp. Tumor Res. 15*, 1-75.

3. Bednar, T. W. and E. M. Linsmaier-Bednar (1971). Induction of cytokinin-independent tobacco tissues by substituted fluorenes. *Proc. Nat. Acad. Sci. U.S. 68*, 1178-1179.

4. Binns, A. and F. Meins, Jr. (1973). Evidence that habituation of tobacco pith cells for cell division promoting factors is heritable and potentially reversible. *Proc. Nat. Acad. Sci. U.S., 70*, 2660-2662.

5. Braun, A. C. (1947). Thermal studies on the factors responsible for tumor initiation in crown-gall. *Am. J. Bot. 34*, 234-240.

6. Braun, A. C. (1953). Bacterial and host factors concerned in determining tumor morphology in crown-gall. *Bot. Gaz. 114*, 363-371.

7. Braun, A. C. (1958). A physiological basis for autonomous growth of the crown-gall tumor cell. *Proc. Nat. Acad. Sci. U.S. 44*, 344-349.

8. Braun, A. C. (1959). A demonstration of the recovery of the crown-gall tumor cell with the use of complex tumors of single-cell origin. *Proc. Nat. Acad. Sci. U.S. 45*, 932-938.

9. Braun, A. C. (1961). Plant tumors as an experimental model. *Harvey Lectures 56*, 191-210.

10. Braun, A. C. (1969). "The Cancer Problem. A Critical Analysis and Modern Synthesis." Columbia University Press, New York.

11. Braun, A. C. ed. (1972a). "Plant Tumor Research." S. Karger, Basel.

12. Braun, A. C. (1972b.) The relevance of plant tumor systems to an understanding of the basic cellular mechanisms underlying tumorigenesis. *Progr. Exp. Tumor Res. 15*, 165-187.

13. Braun, A. C. (1974). The cell cycle and tumorigenesis in plants. In "The Cell Cycle and Cell Differentiation," H. Holtzer and J. Reinert, eds., Springer-Verlag (in press).

14. Braun, A. C. and R. J. Mandle (1948). Studies on the inactivation of the tumor-inducing principle in crown-gall. *Growth 12*, 255-269.

15. Braun, A. C. and P. R. White (1943). Bacteriological sterility of tissues derived from secondary crown-gall tumors. *Phytopathology 33*, 85-100.

16. Brink, R. A., E. D. Styles, and J. D. Axtell (1968). Paramutation: directed genetic change. *Science 159*, 161-170.

17. Cahn, R. D. and M. B. Cahn (1966). Heritability of cellular differentiation in retinal pigment cells *in vitro*. *Proc. Nat. Acad. Sci. U.S. 55*, 106-114.

18. Camus, G. and R. J. Gautheret (1948). Sur la transmission par greffage des propriétés tumorales des tissus de crown-gall. *C.R. Soc. Biol. 142*, 15-16.

19. Chadka, K. C. and B. I. S. Srivastava (1971). Evidence for the presence of bacteria-specific proteins in sterile crown-gall tumor tissue. *Plant Physiol. 48*, 125-129.

20. Coon, H. G. (1966). Clonal stability and phenotypic expression of chick cartilage cells *in vitro*. *Proc. Nat. Acad. Sci. U.S. 55*, 66-73.

21. Davidson, E. H. (1968). "Gene Activity in Early Development." Academic Press, New York.

22. Delbrück, M. (1949). In discussion following paper by Sonneborn and Beale. *Colloq. Int. Centre Nat. Recherche Sci., Paris 7*: 25.

23. deRopp, R. S. The growth-promoting and tumefacient factors of bacteria-free crown-gall tumor tissue. (1947). *Am. J. Bot. 34*, 248-261.

24. Duffield, E. C. S., S. D. Waaland, and R. Cleland (1972). Morphogenesis in the red alga *Griffithsia pacifica*: regeneration from single cells. *Planta 105*, 185-195.

25. Dyson, W. H. and R. H. Hall (1972). $N^6-(\Delta^2$-isopentenyl) adenosine: its occurrence as a free nucleoside in an autonomous strain of tobacco tissue. *Plant Physiol. 50*, 616-621.

26. Eguchi, G. and T. S. Okada (1973). Differentiation of lens tissue from the progeny of chick retinal pigment cells cultured *in vitro*: a demonstration of a switch of cell types in clonal cell culture. *Proc. Nat. Acad. Sci. U.S. 70*, 1495-1499.

27. Einset, J. W. and F. Skoog (1973). Biosynthesis of cytokinin in cytokinin-autotrophic tobacco callus. *Proc. Nat. Acad. Sci. U.S. 70*, 658-660.

28. Fox, J. E. (1963). Growth factor requirements and chromosome number in tobacco tissue cultures. *Physiol. Plant. 16*, 793-803.

29. Gautheret, R. J. (1955). The nutrition of plant tissue cultures. *Ann. Rev. Plant Physiol. 6*, 433-484.

30. Gehring, W. (1968). The stability of the determined state in cultures of imaginal disks in *Drosophila*. *In*: "The Stability of the Differentiated State" (H. Ursprung, ed.) pp. 134-154. Springer, Berlin.

31. Gurdon, J. B. and V. Uehlinger (1966). "Fertile" intestine nuclei. *Nature 210*, 1240-1241.

32. Jablonski, J. R. and F. Skoog (1954). Cell enlargement and cell division in excised tobacco pith tissue. *Physiol. Plant. 7*, 16-24.

33. Kacser, H. (1963). The kinetic structure of organisms. *In*: "Biological Organization at the Cellular and Supercellular Level" (H. Harris, ed.) pp. 25-41. Academic Press, New York.

34. Limasset, P. and R. Gautheret (1950). Sur le caractère tumoral des tissus de tabac agant subi le phénomène d'accoutumanace aux hétéro-auxines. *C.R. Acad. Sci., Paris, 230*, 2043-2045.

35. Lutz, A. (1971). Aptitudes morphogénétiques des cultures de tissus d'origine unicellulaire. *In*: "Les Cultures de Tissus de Plantes," *Colloq. Int. Cent. Nat. Rech. Scient.* No. 193, pp. 163-168. Editions du Centre Nat. Rech. Scient., Paris.

36. Meins, F., Jr. (1969). Control of phenotypic expression in tobacco teratoma tissues. Ph.D. Diss., Rockefeller University, New York.

37. Meins, F., Jr. (1971). Regulation of Phenotypic expression in crown-gall teratoma tissues of tobacco. *Develop. Biol. 24*, 287-300.

38. Meins, F., Jr. (1972). Stability of the tumor phenotype in crown-gall tumors of tobacco. *Progr. Exp. Tumor Res. 15*, 93-109.

39. Meins, F., Jr. (1973). Evidence for the presence of a readily transmissible oncogenic principle in crown-gall teratoma cells of tobacco. *Differentiation 1*, 21-25.

40. Miller, C. O., F. Skoog, M. H. Von Saltza, and F. M. Strong (1955). Kinetin, a cell division factor from deoxyribonucleic acid. *J. Am. Chem. Soc. 77*, 1392.

41. Milo, G. E. and B. I. S. Srivastava (1969). RNA-DNA hybridization studies with crown-gall bacteria and the tobacco tumor tissue. *Biochem. Biophys. Res. Comm. 34*, 196-199.

42. Monod, J. and F. Jacob (1961). General conclusions: telenomic mechanisms in cellular metabolism, growth, and differentiation. *Cold Spr. Harb. Symp. Quant. Biol. 26*, 389-401.

43. Murashige, T. and R. Nakano (1966). Tissue culture as a potential tool in obtaining polyploid plants. *J. Heredity 57*, 115-118.

44. Pitot, H. C. and C. Heidelberger (1963). Metabolic regulatory circuits and carcinogenesis. *Cancer Res. 23*, 1694-1700.

45. Quétier, F., T. Huguet, and E. Guille (1969). Induction of crown-gall: partial homology between tumor-cell DNA, bacterial DNA, and the G + C rich DNA of stressed normal cells. *Biochem. Biophys. Res. Comm. 34*, 128-133.

46. Sacristán, M. D. (1967). Auxin-autotrophy and Chromosomenzahl. *Mol. Gen. Genetics 99*, 311-321.

47. Sacristán, M. D. and G. Melchers (1969). The caryological analysis, of plants regenerated from tumorous and other callus cultures of tobacco. *Mol. Gen. Genetics 105*, 317-333.

48. Sacristán, M. D. and M. F. Wendt-Gallitelli (1971). Transformation of auxin-autotrophy and its reversibility in a mutant line of *Crepis capillaries* callus culture. *Mol. Gen. Genetics 110*, 355-360.

49. Schaeffer, G. W. and H. H. Smith (1963). Auxin-kinetin interaction in tissue cultures of nicotiana species and tumor conditioned hybrids. *Plant Physiol. 38*, 291-297.

50. Schildperoort, R. A., H. Veldstra, S. O. Warnaar, G. Mulder and J. S. Cohen (1967). Formation of complexes between DNA isolated from tobacco crown-gall tumors and RNA complementary to *Agrobacterium tumefaciens* DNA. *Biochem. Biophys. Acta 145*, 523-525.

51. Skoog, F. and C. O. Miller (1957). Chemical regulation of growth and organ formation in plant tissues cultured *in vitro. Soc. Exp. Biol. Symp. 11*, 118-131.

52. Smith, H. H. (1972). Plant genetic tumors. *Progr. Exp. Tumor Res. 15*, 138-164.

53. Smith, H. H. and S. A. Sand (1957). Genetic studies on somatic instability in cultures derived from hybrids between *Nicotiana langsdorffii* and *N. sanderae. Genetics 42*, 560-582.

54. Takebe, I., G. Labib and G. Melchers (1971). Regeneration of whole plants from isolated mesophyll protoplasts of tobacco. *Naturwissenschaft 58*, 318-320.

55. Vasil, V. and A. C. Hildebrandt (1965). Differentiation of tobacco plants from single, isolated cells in microculture. *Science 150*, 889-892.

56. Wood, H. N., A. C. Braun, H. Brandes, and H. Kende (1969). Studies on the distribution and properties of a new class of cell division-promoting substances from higher plant species. *Proc. Nat. Acad. Sci. U.S. 62*, 349-356.

57. Wood, H. N., M. C. Lin, and A. C. Braun (1972). The inhibition of plant and animal adenosine 3 ':5 ' — cyclic monophosphate phosphodiesterases by a cell-division-promoting substance from tissues of higher plant species. *Proc. Nat. Acad. Sci. U.S. 69*, 403-406.

SESSION II
Cell Profileration, Differentiation and Neoplasia

Moderator: Brian Spooner

CELL CONTACT AND CELL DIVISION

LaRoy N. Castor

The Institute for Cancer Research
Fox Chase Center for Cancer and Medical Sciences
Philadelphia, Pennsylvania 19111

In cell cultures from normal, non-neoplastic tissue, the rate of cell division generally decreases as the cells become more crowded. This *density-dependent inhibition*[1] (Stoker and Rubin, 1967) is reversible, as shown by "wound-healing" experiments and by a return to maximum division rates after subcultivation (Todaro *et al.*, 1964; Castor, 1968; Dulbecco and Stoker, 1970). However, density-dependent inhibition is absent or much reduced in cultures of neoplastic cells, which tend to continue dividing at close to their maximum rate until they become necrotic.

Efforts to establish the cause of density-dependent inhibition have led to much controversy about the relative influence of contacts between cells and factors in the culture medium. Without doubt, the rate of cell division is strongly affected by the condition of the medium, which in turn varies with the number of cells and with the time since the medium was last changed. Division becomes partially inhibited as lactic acid is produced and the pH decreases (Ceccarini and Eagle, 1971; Rubin, 1971). Cells also produce other inhibitors (Burk, 1966; Garcia-Giralt *et al.*, 1970), and some cells produce factors that stimulate division (Shodel, 1972; Dulak and Temin, 1973). Of the ingredients supplied with fresh medium, the best-recognized division-stimulating factors are serum macromolecules (Paul *et al.*, 1971; Clarke and Stoker, 1971; Temin *et al.*, 1972). The addition of fresh serum to a density-inhibited culture is followed about a day later by a burst of cell division. By this time the cells have depleted the medium of the serum factors, and division continues at a reduced rate or ceases entirely until fresh serum is again added (Todaro *et al.*, 1965; Holley and Kiernan, 1968). Other growth-stimulating ingredients of the medium, such as glucose, glutamine or

---

[1] or *cell cycle inhibition* (Marciera-Coelho, 1967). The inhibition as it occurs in 3T3 cells was termed *contact inhibition of cell division* by the original investigators (Todaro *et al.*, 1964), *topoinhibition* by Dulbecco (1970) and *post-confluence inhibition* by Martz and Steinberg (1972).

one or more amino acids, may also be depleted (Blaker *et al.*, 1971; Griffiths, 1972).

For definitive study of the population kinetics of density-dependent inhibition, all these factors can be maintained at essentially fixed concentrations by the continuous perfusion of fresh medium through the culture (Kruse *et al.*, 1969). If in addition the culture is monitored by time-lapse cinemicrography, effects of contacts between cells in initiating density-dependent inhibition become readily apparent (Castor, 1968). This paper is concerned with the contact interactions, how they lead to control of division in cells cultured from normal tissues, and how the control mechanism breaks down in tumor cells.

METHODS AND DEFINITIONS

The perfusion method we have used (Fig. 1) is designed to feed a culture in a 35 mm dish continuously for a week or more, and it also provides for continuous stirring of the culture medium (Castor, 1973). A typical rate of perfusion for a confluent culture is 1 ml/day per cm^2 of culture surface. Lower rates may be necessary to avoid necrosis in cultures containing fewer cells, especially if the medium contains low concentrations of serum. Higher rates are used in extremely crowded cultures.

The cinemicrographic measurements of population kinetics to be presented here were made through an inverted microscope with a low power objective (4X or 10X) focused directly on the film plane of the 16 mm camera. The kinetics were determined by frame-by-frame inspection of the

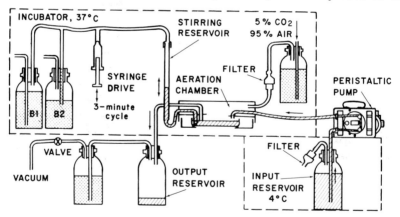

Fig. 1. *Diagram of perfusion apparatus. Culture medium contained in a refrigerated input reservoir is fed to the culture dish by a peristaltic pump and removed by vacuum aspiration. Stirring of the medium in the dish is provided by the syringe with reciprocating drive and the pressure-regulating bubblers, B1 and B2. (From Castor, 1973).*

film in an analyzer projector. Rates of cell division and movement were plotted as functions of population density, so as to provide characteristic curves that are distinctive for a particular cell type, independent of time and of special culture conditions such as the starting inoculum. *Mitotic density* is the number of mitoses counted *per unit area* per unit time. It is essentially the slope of the usual growth curve and is therefore a sensitive indicator of the rate of culture growth. *Population density* of pre-confluent cultures was calculated directly from counts of cells in a defined area of the microscope field. After confluence, the population density was estimated from the pre-confluent count plus the total number of mitoses observed up to the given time. For density-inhibited cells this procedure gave good agreement with counts of nuclei in cultures that were fixed at the close of the experiment. For neoplastic cells having high death rates the error was not too great over the range of population density of interest. *Confluence*, indicated by arrows on the charts, is the lowest population density at which all the culture surface within the field of view is covered with cells. At confluence almost every cell is in contact with other cells around its entire border, as shown by the strong tendency of the cells to avoid overlapping each other. This purely optional definition of "contact" is the only one implied here. *Average motility* is the average of the distances traveled per unit time, in all directions, by a representative number of cells.

POPULATION KINETICS OF DENSITY-INHIBITED CELLS

Contact Inhibition of Division of 3T3 and Epithelial-like Cells

The first evidence that contacts between cells might influence the rate of cell division came from experiments of Todaro and Green (1963) with a mouse embryo cell line designated 3T3 to signify a sub-cultivation procedure that maintained the cells at non-confluent population densities. Cultures not sub-cultivated ceased growing at a low population density, called the saturation density. However, the saturation density depended strongly on the concentration of serum in the medium, and with conventional culture methods it was not possible to decide how much of the inhibition resulted from depletion of serum factors and how much from intercellular contacts (Todaro *et al.*, 1967; Holley and Kiernan, 1968).

The population kinetics of 3T3 cells in perfusion culture (Fig. 2) demonstrated that intercellular contact was a primary determinant of inhibition of division (Castor, 1971). In Fig. 2 the mitotic density and average motility are plotted as functions of population density. Since the rate of division of an exponentially growing culture is directly proportional to the number of cells present, exponential growth is indicated as a diagonal straight line. Onset of inhibition of division is indicated by the departure of the

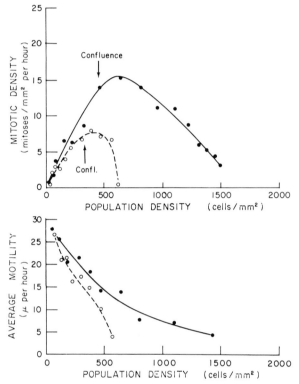

Fig. 2. *Contact inhibition of cell division and motility, illustrated by characteristic curves of mitotic density (mitoses per unit area per unit time) and average motility (average speed in any direction) vs. population density (cells per unit area). 3T3 cells, strain 42L.* ●———●, *30% dialyzed calf serum;* ○———○, *10% dialyzed calf serum. (From Castor, 1971).*

mitotic density curve from linearity, beginning at confluence (arrows) in both concentrations of serum. The cells in 30% serum were slightly less flattened on the culture surface than those in 10% serum, and therefore confluence occurred at a slightly higher population density.

Prior to confluence the mitotic density in 10% serum was at least two-thirds of the mitotic density in 30% serum, for any population density, showing that serum concentration had little effect on exponential growth rates of non-confluent cultures. However, after confluence the higher serum concentration stimulated a higher rate of division, thus antagonizing the inhibitory effects of population density and becoming the major determinant of the saturation density. At the saturation density cell division ceased; and no significant amount of cell death occurred during the next 3 days that were monitored by cinemicrography. This behavior contrasts with that of saturated 3T3 cultures fed 2 or 3 times a week, in which the mitotic activity after each feeding was equalized by cell death prior to the next feeding (Todaro *et al.*, 1965). Evidently, a given serum concentration supports a rigorously defined population density of 3T3 cells, and any changes in the effective serum concentration causes a compensating change in the population density.

The lower plot in Fig. 2 shows that the average motility decreased steadily as the population increased. The decrease began well before confluence, so there was no obvious relationship between motility and division rate in non-confluent cultures. The relationship that is significant is that after confluence the motility and the mitotic density decreased together, maintaining an approximate proportionality to each other. This proportionality after confluence occurs for all the density-inhibited lines we have investigated, within limitations of experimental error that become rather wide at low values of motility.

When 3T3 cells are maintained at their saturation density for most of the time between subcultivations, the attainable saturation density gradually increases over the months, and the cells become comparable to the 3T6 and 3T12 lines derived by Todaro and Green (1963). Contact inhibition of cell division is retained in these "spontaneously transformed" cells (Fig. 3). The significant change in growth characteristics is a decrease of about 4-fold in the serum requirement for a given rate of division at a given population density.

It should be clear from these results that the frequently expressed notion that a low saturation density indicates contact inhibition of division, while a high saturation density indicates lack of contact-inhibition, has questionable usefulness. The argument as it applies to high saturation density is specifically contradicted in Figs. 2 and 3, in which inhibition begins at confluence even in cultures that eventually reach a high saturation density. The argument as it applies to low saturation density has some validity, provided there is a demonstrable inhibition of division, and not simply a high rate of cell death. Failure of some "flat" variants of virally-transformed cells to satisfy this criterion has been demonstrated by Dulbecco (1970) and Scher and Nelson-Rees (1971).

Epithelial-like rat liver cells also exhibit contact inhibition of cell division (Fig. 4). These diploid cells are smaller than the heteroploid 3T3 cells, so that their population density at confluence was higher, but nevertheless their mitotic density curve departed from linearity shortly before confluence, and after confluence the mitotic density and motility decreased approximately in proportion. Division did not cease entirely. The lowest mitotic density corresponds to about a 15-fold increase in doubling time. Unlike 3T3 cells, when the inhibition in these liver cells was maximal there was slow turnover in the population due to a small rate of cell death.

The changing morphology of a contact-inhibited culture as the population density increased is illustrated for the rat liver cells (Fig. 5). When growth was exponential prior to confluence the cells were well flattened. They made

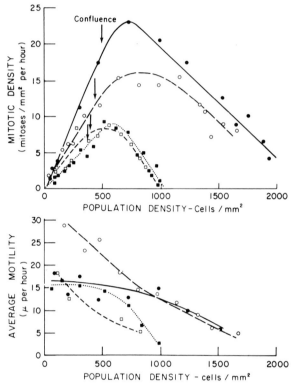

Fig. 3. *Contact inhibition of cell division and motility in "spontaneously trans-formed" 3T3 cells, strain 42H, showing a much decreased requirement for serum compared to strain 42L in Fig. 2.* ●——●, *30% dialyzed calf serum;* ○——○, *10% dialyzed calf serum;* □——□, *3% dialyzed calf serum;* ■——■, *1% dialyzed calf serum. (From Castor, 1971).*

many contacts with each other; but because of the continued motility these contacts were transient, giving the culture a loose epithelial-like appearance. No measurable inhibition of division was associated with the transient contacts. Inhibition began at confluence (center picture), coincidental with the beginning of a change in cell shape to a more compact form. The increasing compactness was enforced by strong contact inhibition of the movement of cells across each other's surfaces (Abercrombie and Heaysman, 1954). In post-confluent cultures (Fig. 5, right) there were localized differences in the saturation density. We have not investigated the reasons for this, but the shapes and division rates of clones in different local areas would not necessarily respond identically to crowding.

The monolayer of compact, polygonal rat liver cells of Fig. 5 (right) bears at least a superficial resemblance to the cords of hepatocytes in the intact

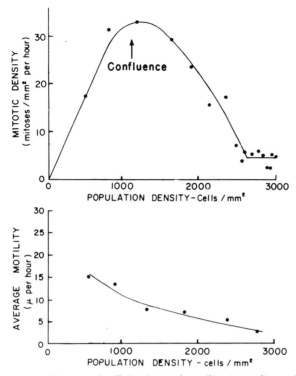

Fig. 4. *Contact inhibition of cell division and motility in rat liver cells, strain 3C4 (Coon, 1968), in a medium containing 5% fetal bovine serum.*

liver. Like many glandular tissues, the liver has a 3-dimensional structure that can be viewed as an assembly of monolayers — in which the parenchymal cells contact each other along their borders, but have dorsal and ventral surfaces that lie along sinusoids. Since liver cells in culture attain an essentially similar arrangement, they may possibly express whatever features of the growth control mechanism depend upon arrangement. There is thus good reason to suppose that contact-inhibitory control of cell division in culture is a reflection of a basic mechanism of growth control *in vivo*. However, *in vivo* there are also other control mechanisms not found in culture, such as a specificity of attachment of homologous cells to each other, and hormonal influences.

Contact Regulation of Division of Epithelial-like Cells

In a second class of epithelial-like cells the control of division is described as "contact regulation" because the mitotic density is regulated to a constant value in post-confluent cultures (Castor, 1968). This is illustrated for the 1S1

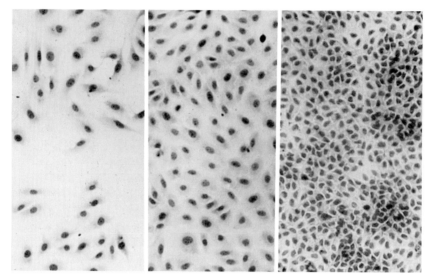

Fig. 5. *The rat liver cells of Fig. 4 at pre-confluent, confluent, and post-confluent population densities. No appreciable change in cell shape occurred until after confluence, when the combined effects of contact inhibition of movement and the increasing population density forced the cells into more compact shapes. Hematoxylin and eosin. 120 X.*

line from mouse spleen (Fig. 6). A similar constancy in the rate of cell increase per unit area was found by Zetterberg and Auer (1970) in epithelial cells from mouse kidney in primary culture. Motility decreased to a low value in Fig. 6 even before confluence. The low motility probably resulted from the adhesiveness of the cells, which formed into tight epithelial-like "islands" (Fig. 8, left).

A constant mitotic density implies that the rate of division *per cell* decreased directly in proportion to the average area of culture surface occupied per cell. The average area decreased steadily as the strong contact inhibition of movement changed the shapes of the cells from highly flattened to highly compact (Fig. 8). Eventually, the increasing population forced the compact cells to overlap and to form multiple layers. Counts of mitotic figures in these layers showed that most of them were in the lowest layer, next to the culture surface (Fig. 7). Thus, division in the upper layers was strongly inhibited, while division in the lower layer continued at a rate proportional to surface area.

Contact-regulated growth is analogous to that of multiple-layered epithelium, such as epidermis, *in vivo*. In epidermis the zone of active mitosis is close to the basal layer, and we might assume that the mitoses are stimulated by diffusion of some factor from the dermis. But human epidermis, freed of

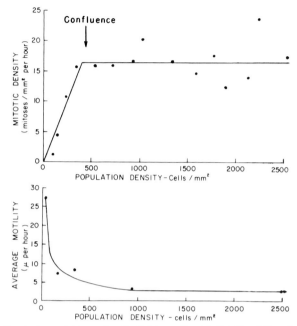

Fig. 6. *Contact regulation of cell division of epithelial-like 1S1 cells, in a medium containing 20% calf serum. After confluence the mitotic density remained approximately constant, indicating a rate of division proportional to the average area occupied per cell on the culture dish. Motility became almost completely inhibited when epithelial-like "islands" formed shortly after the culture was prepared. (Re-plotted from Castor, 1968, 1970).*

dermis and placed on an artificial substrate in culture, maintains the same distribution. Dividing cells are found closest to the substrate and non-dividing cells in the upper layers, which in culture are closest to the nutrient supply (Flaxman *et al.*, 1967). Cultures of this type have been maintained for 3 to 4 months, with mitoses continuing in the lower layers and with the upper layers of keratinized cells becoming steadily thicker (B. A. Flaxman, personal communication). Attachment to a substrate appears to be the stimulant to division.

Partial Inhibition of Division of Fibroblasts

WI-38 fibroblasts were found to exhibit density-dependent inhibition prior to confluence (Fig. 9), when there was still considerable empty space between the cells on the culture surface. However, a change in cell shape also occurred prior to confluence (Fig. 11). Although at low densities the cells were stellate or only slightly elongated, as shown at the left of the figure, their elongation increased as they became constrained into an orientation parallel to their axes (Fig. 11, center and right).

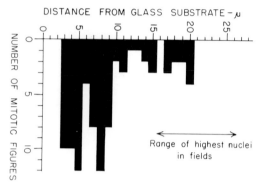

Fig. 7. *Distribution of heights of mitotic figures above the glass at the end of the experiment of Fig. 6, showing strong inhibition of division of cells not close to the culture surface. (From Castor, 1968).*

Beyond confluence the mitotic density and the motility (Fig. 9, lower) both decreased with increasing population density, but only gradually. In the motility plot of Fig. 9, the dashed line is intended to indicate that cell movement continued to high population densities. The average motility could not be accurately measured in the post confluent culture because most individual cells were not separately identifiable. However, the time-lapse films showed movement of an undetermined proportion of cells along the axis of orientation.

Serum protein influenced the division of WI-38 cells comparably to its influence on 3T3 cells, i.e., it affected mainly the *inhibition* (Fig. 10). Pre-confluent, non-inhibited cells were relatively insensitive to serum, except that the lower the serum concentration the higher the population density required for growth to begin. The results of Fig. 10 were obtained with the cells growing on small pieces of cover slip in a larger dish, with the medium replaced almost completely by a rapid 2-hour perfusion each day. This departure from continuous perfusion was necessitated by a high sensitivity of WI-38 to the washing-out of conditioning factors (Rein and Rubin, 1971) when the medium contains low concentrations of serum. Although the post-confluent mitotic density in 10% serum was lower than for the corresponding population density with continuous perfusion, the dependence of mitotic density on serum concentration is clearly evident.

Baker and Humphreys (1971) have demonstrated in chick fibroblasts a causal relationship between serum concentration on the one hand and motility and synthesis of DNA on the other. Confluent cultures in 0.5% serum were stimulated by additional serum. Within minutes the rate of cell movement had increased, and the cells became less well spread on the culture dish. Over the range of serum concentrations from 0.5% to 20% the increase in rate of cell movement was quantitatively proportional to the number of cells stimulated

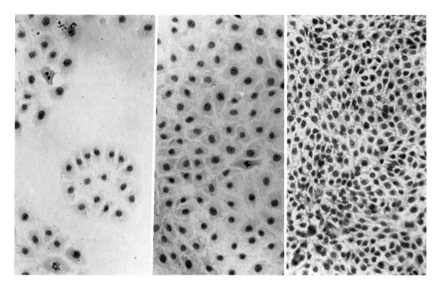

Fig. 8. *The epithelial-like 1S1 cells of Figs. 6 and 7 at pre-confluent, confluent, and post-confluent population densities. These cells differed from those of Fig. 5 in attaching to each other more firmly, so that motility was low even in non-confluent cultures. Hematoxylin and eosin. 120 X.*

to synthesize DNA, suggesting that both movements and growth were released by a common serum activity.

For the two types of density-dependent inhibition considered previously, we were able to speculate that a comparable inhibition could occur *in vivo*. However, for cultured fibroblasts, the inhibition is at best just a partial manifestation of an *in vivo* control mechanism. *In vivo* the cells exist in a 3-dimensional collagenous matrix, which can provide physical and chemical interactions that are difficult to reproduce in culture (Bard and Elsdale, 1971). Cell shape and cell movement, for example, can be controlled in all three dimensions. In culture the cells become highly elongated and at least some of them continue to move along their axes. Since inhibition of movement is associated with inhibition of division, it is not surprising that division is only partially inhibited at high population densities. This lack of axial restraint and the formation of dense cell sheets has no parallel in normal fibroblasts *in vivo*.

NEOPLASTIC GROWTH IN CULTURE

Mitotic density and motility characteristics for a tumor cell line in perfusion culture are illustrated in Fig. 12. This line, 1S13, was derived from a subcutaneous tumor formed in a mouse by the 1S1 cells of Figs. 6-8. Cell

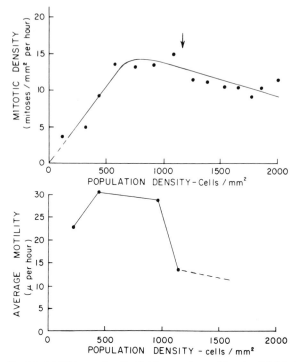

Fig. 9. *Partial inhibition of cell division and motility in WI-38 fibroblasts. The inhibition began at about 600 cells/mm² – only half the population density of confluence (arrow). Average motility decreased at confluence. Some cells continued to move even at high population densities, but accurate measurements of motility at high densities were not possible.*

flattening is perceptibly reduced compared to non-tumor cells of the same intracellular volume, and in the same concentration of serum. Confluence therefore occurs at a higher population density than it did for 1S1 cells. Nevertheless, there is no inhibition of division at confluence, but exponential growth continues for one or more additional doublings of the population. Likewise, there is no pronounced decrease of motility with the increasing population; and in fact in one neoplastic line (HeLa) motility actually increases with population density (Castor, 1970).

Flattening of the neoplastic cells increases, and density-dependent inhibition of division and motility can be restored to varying degrees by reduction of the concentration of serum, as illustrated in Fig. 12 for 1% serum. Thus, susceptibility to density-dependent inhibition does not of itself distinguish cultured normal cells from cultured neoplastic cells, although this is the usual consequence of growing them in 10% serum. A more fundamental indicator is the decreased serum requirement, determined quantitatively as the concentra-

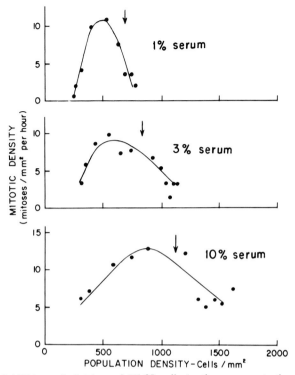

Fig. 10. *Inhibition of division of WI-38 cells in three concentrations of dialyzed serum. Confluence (arrows) occurred at higher population densities in the higher concentrations of serum, reflecting a decreased degree of cell flattening. Because of differences in technique, the values of mitotic density are not comparable with Fig. 9.*

tion of serum required to produce a given mitotic density at some population density beyond confluence.

The evolution toward a lower serum requirement may be a general characteristic of heteroploid cell lines in culture, and this tendency may explain why continued cultivation leads to cells with neoplastic capability. Fig. 3 has illustrated the population kinetics of 3T3 cells that had "transformed spontaneously" to a lower serum requirement. These cells were from a non-inbred mouse strain, so they cannot be implanted and tested directly for neoplastic potential. But comparable lines of 3T6 and 3T12 cells from BALB/c tissue are capable of forming tumors, while 3T3 cells from the same tissue are not (Aaronson and Todaro, 1968).

An even lower serum requirement is exhibited by 3T3 cells transformed by the oncogenic virus, SV40 (Fig. 13). Cultures in 10% serum grew rapidly and formed into irregular clumps, preventing the determination of an accurate mitotic density curve. But even when the serum concentration was reduced to

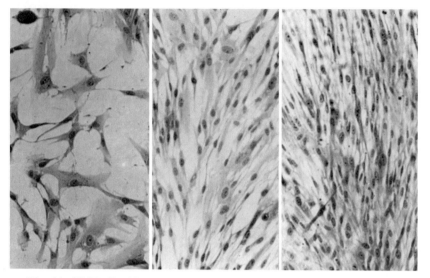

Fig. 11. *WI-38 fibroblasts at two pre-confluent densities and after confluence, in medium containing 10% calf serum. An increase in cell elongation occurred prior to confluence in association with the onset of inhibition of division shown in Fig. 9. Hematoxylin. 120 X.*

1% or 0.3%, contact-inhibited growth like that of the antecedent 3T3 cells did not occur. However, proliferation was slower than in 10% serum; and the decreasing motility suggested that contact interactions had been restored (Castor, 1971).

NATURE OF THE INHIBITORY SIGNAL

The earliest hypotheses of the nature of contact inhibition of cell division attempted to relate it to the cell membrane interactions that prevent cells from moving across each other — the contact inhibition of cell movement first described by Abercrombie and Heaysman (1954, Abercrombie, 1970; Todaro *et al.*, 1964; Eagle, 1965). However, even many tumor cells do not readily overlap each other, although they may cross over normal cells (Veselý and Weiss, 1973). Contact inhibition of movement occurs in non-confluent cultures when only parts of cells come into contact, but there is no inhibition of division. Martz and Steinberg (1972) did not detect any increased generation times even in those 3T3 cells in pre-confluent culture that happened to remain in all-around contact through the entire period between two successive mitoses. Thus, contact inhibition of movement is not itself sufficient to cause density-dependent inhibition of division. On the other hand, contact inhibition

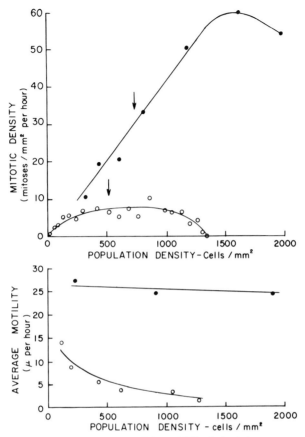

Fig. 12. *Population kinetics of a cell line, 1S13, derived from a tumor formed by the 1S1 line of Figs. 6-8. ●————●: With 20% calf serum in the culture medium, neither cell division nor motility was controlled. ○————○: Density-dependent inhibition was restored when the concentration of calf serum was lowered to 1%. (Data for 20% serum re-plotted from Castor, 1968).*

of movement is probably necessary for inhibition of division, since it constrains the cells to change their shape rather than to overlap. It was change in shape that was correlated with the beginning of inhibition, either at confluence (for contact inhibition and contact regulation) or prior to confluence (for fibroblasts).

How can a restriction of cell shape inhibit division? Work in Stoker's laboratory demonstrates that cells showing density-dependent inhibition are also anchorage-dependent, i.e., they require attachment to a substrate such as glass, plastic, or collagen in order to divide (Stoker *et al.*, 1968). For the saturation density of 600 cells/mm^2 of 3T3 cells, we would expect an average

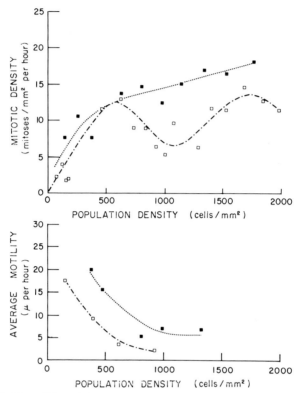

Fig. 13. *Very slight inhibition of division in cultures of SV-transformed 3T3 cells, even with low concentrations of serum in the culture medium. The inhibition of motility was comparable to that for the non-transformed cells in higher concentrations of serum (Fig. 2).* ■——■, *1% dialyzed calf serum;* □——□, *0.3% dialyzed calf serum. (From Castor, 1971).*

cell to require greater than 1600 μm^2 of substrate area, or about 46 μm in perpendicular linear directions. Maroudas (1972, 1973) has grown 3T3 cells on small platelets and has confirmed that these approximate dimensions are needed if the cells are to divide. However, he also finds that cells will divide attached to glass fibers of 30 μm minimum length (Fig. 14), although these have a surface area of only about 60 μm^2. The conclusion is that only one linear dimension is required for anchorage, and he downgrades the necessity for a specific area over which anchorage is required. The working hypothesis is suggested that anchorage causes tension or deformation of the cell membrane, which in turn releases an internal chemical transmitter that induces cell division. We can imagine that in mass cultures of density-inhibited cells the restriction imposed by crowding would prevent the cell surface from stretching and deforming, and therefore cell division would be inhibited.

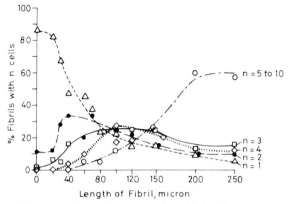

Fig. 14. *Proliferation of 3T3 cells on glass fibrils. Cells were seeded on fibrils suspended in agar, and after 6 days the percent of fibrils having n cells attached was determined. Values of n greater than unity are evidence that the cells have divided. (From Maroudas, 1973).*

In addition to the change in cell shape, an increasing inhibition of division was persistently correlated in our experiments with a decreasing motility. Decreasing motility probably reflects physical adhesions between cell borders (Abercrombie and Heaysman, 1953; Castor, 1970; Baker and Humphreys, 1971), although motility could also be controlled by intracellular processes that are affected by contact interactions. In either case, movement of the cell requires movements of its surface (Ingram, 1969), so that motility measurements provide an indirect measure of the physical activity of the cell surface. Since serum protein stimulates motility and also cellular synthetic processes that culminate in cell division, it is not unlikely that surface activity promotes the pinocytosis of macromolecules from serum (Castor, 1971, 1972).

Differences in pinocytotic activity of normal and tumor cells are seen in the scanning electron microscope. Not only are tumor cell surfaces more active, but the activity occurs over the entire cell surface. In contrast, the pinocytotic activity and cytoplasmic extrusions of normal cells are confined to the periphery and the lower surface, a finding consistent with our previous discussion of anchorage dependence (Westermark *et al.*, 1972; Veselý and Boyde, 1973). Veselý (1972) examined the surface activity of living cells by high resolution phase-contrast cinemicrography. The dorsal surfaces of the tumor cells had microfolds, microvilli, and pinocytotic orifices that appeared to be continuously changing their shape and, to a slight degree, their locations (Fig. 15). In normal rat fibroblasts the dorsal surfaces were only slightly wrinkled, with virtually no evidence of any surface movements (Fig. 16). Pinocytosis was confined to the broad lamellipodia at the leading edge (not shown). Veselý and Weiss (1973) have also demonstrated that pinocytosis increases in density-inhibited cells near the edge of a wound.

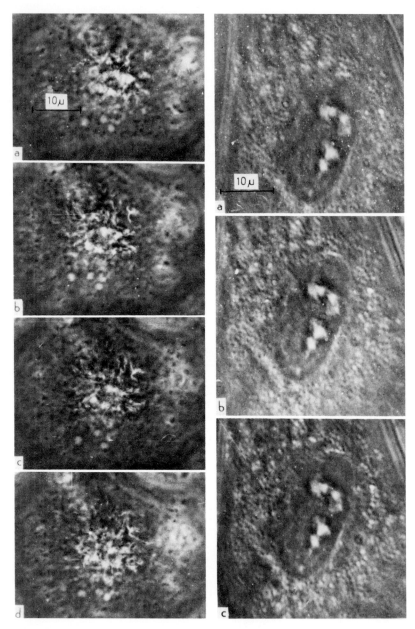

Fig. 15. *Part of the dorsal surface of an epithelial-like tumor cell, RsK4, showing the motility of surface structures. Pictures were taken at consecutive intervals of 10 seconds by high-resolution negative phase contrast cinemicrography. (From Veselý, 1972).*

Fig. 16. *Part of the dorsal surface of a rat fibroblast, LWF. The slight wrinkling is due in part to projections of internal cell structures. There is no noticeable motility of the surface projections. Consecutive intervals of 15 seconds. (From Veselý, 1972).*

Little concrete evidence exists about how cell surface stretching or pinocytosis could stimulate cell division, or how macromolecules in serum act on the surface or within the cell. One problem is specificity: Do the particular macromolecular factors fractionated from serum have a particular ability to induce cell surface activity, or is the activity induced by a broader spectrum of macromolecules and the specific molecule pinocytosed and used intracellularly? In either case, the pinocytosis could increase the net transfer of small molecules as well as large molecules. Zinc, phosphate, amino acids, or combinations of several small molecules have been proposed as being stimulators of intracellular synthetic processes and also being limited in their transport in density-inhibited cells (Griffiths, 1972; Holley, 1972, Rubin and Koide, 1973). However, no investigator has yet duplicated with small molecules alone the ability of serum macromolecules to promote cell division at high population densities (Temin et al., 1972).

Neoplastic cells lack the ability to restrict each other's surface movements, as shown by their continued motility in crowded cultures. They also have a reduced serum requirement, for the same rate of division at the same population density. It is not clear which, if either, of these properties is secondary to the other. One hypothesis would be that the primary lesion in neoplasia is at the cell surface, which admits into the cell an excess of nutrients and factors needed for growth and for continued stimulation of the surface movements. An alternative hypothesis would be that the intracellular machinery responsible for growth and movement has a reduced need for factors from outside the cell.

These considerations lead to two pairs of questions about growth control and neoplasia: 1) How is the division-stimulatory effect of extracellular macromolecules transmitted from the outside to the inside of the cell? Are there distinctive characteristics of the neoplastic cell surface that prevent it from restricting transmission of the stimulus? 2) On what synthetic or metabolic reactions within the cell does the stimulus act? In the tumor cell are these reactions less dependent on an extracellular stimulus? Some answers appear in other articles in this volume.

SUMMARY

As cultures of cell lines derived from normal, non-neoplastic tissues become crowded, the rate of cell division becomes increasingly inhibited. The inhibition begins concomitantly with a change toward a more compact cell shape imposed by contacts between cells. The inhibition is not caused by depletion or modification of the culture medium by the cells, since it occurs when the medium is continuously perfused. In perfusion culture three types of inhibition can be distinguished by different cell behavior with increasing population density: contact inhibition in 3T3 and some epithelial-like cells; contact regulation in other epithelial-like cells; and a partial inhibition in WI-38 fibroblasts.

The rate of division in post-confluent cultures depends upon the concentration of serum protein in the culture medium. It is proposed that stretching of the cell surface is required for the effect of serum macromolecules to be transmitted into the cell, and that the stretching is increasingly prevented as the population density increases. Neoplastic cells require a lower serum concentration than do normal cells, for a given rate of division at a given population density. Therefore, at a fixed serum concentration they divide more rapidly and reach higher population densities.

ACKNOWLEDGEMENTS

I am grateful for the careful work of Mary Gray, Dolores Cacchio and Josephine Jacobs in analyzing the cinemicrographic films. These studies were supported by U.S.P.H.S. Research Grants CA-07846, CA-10367, CA-06927 and RR-05539 from the National Institutes of Health, and by an appropriation from the Commonwealth of Pennsylvania.

REFERENCES

1. Aaronson, S. A., and Todaro, G. J. (1968). Development of 3T3-like lines from Balb/c mouse embryo cultures: Transformation susceptibility to SV40. *J. Cell. Physiol. 72*, 141-148.

2. Abercrombie, M., and Heaysman, J. E. M. (1953). Observations of the social behavior of cells in tissue culture. I. Speed of movement of chick heart fibroblasts in relation to their mutual contacts. *Exp. Cell Res. 5*, 111-131.

3. Abercrombie, M., and Heaysman, J. E. M. (1954). Observations on the social behavior of cells in tissue culture. II. "Monolayering" of fibroblasts. *Exp. Cell Res. 6*, 293-306.

4. Abercrombie, M. (1970). Contact inhibition in tissue culture. *In Vitro 6*, 128-142.

5. Baker, J. B., and Humphreys, T. (1971). Serum-stimulated release of cell contacts and the initiation of growth in contact-inhibited chick fibroblasts. *Proc. Nat. Acad. Sci. U.S. 68*, 2161-2164.

6. Bard, J., and Elsdale, T. (1971). Specific growth regulation in early subcultures of human diploid fibroblasts. *In* "Growth Control in Cell Cultures," *Ciba Foundation Symp.* (G. E. W. Wolstenholme and J. Knight, eds.), pp. 187-197. Churchill Livingstone, London.

7. Blaker, G. J., Birch, J. R., and Pirt, S. J. (1971). The glucose, insulin and glutamine requirements of suspension cultures of HeLa cells in a defined culture medium. *J. Cell Sci. 9*, 529-537.

8. Bürk, R. R. (1966). Growth inhibitor of hamster fibroblast cells. *Nature 212*, 1261-1262.

9. Castor, L. N. (1968). Contact regulation of cell division in an epithelial-like cell line. *J. Cell. Physiol. 72*, 161-172.

10. Castor, L. N. (1970). Flattening, movement and control of division of epithelial-like cells. *J. Cell. Physiol. 75*, 57-64.

11. Castor, L. N. (1971). Control of division by cell contact and serum concentration in cultures of 3T3 cells. *Exp. Cell Res. 68*, 17-24.

12. Castor, L. N. (1972). Contact inhibitions of cell division and cell movement. *J. Invest. Dermatol. 59*, 27-32.

13. Castor, L. N. (1973). Culture dish perfusion with cinemicrography. *In* "Methods and Applications of Tissue Culture" (P. F. Kruse, Jr. and M. K. Patterson, eds.) Academic Press, New York. pp. 298-303.

14. Ceccarini, C., and Eagle, H. (1971). pH as a determinant of cellular growth and contact inhibition. *Proc. Nat. Acad. Sci. U.S. 68*, 229-233.

15. Clarke, G. D., and Stoker, M. G. P. (1971). Conditions affecting the response of cultured cells to serum. *In* "Growth Control in Cell Cultures," *Ciba Foundation Symp.*, (G. E. W. Wolstenholme and J. Knight, eds.), pp. 17-28.

16. Coon, H. G. (1968). Clonal culture of differentiated rat liver cells. *J. Cell Biol. 39*, 29a.

17. Dulak, N. C., and Temin, H. M. (1973). A partially purified polypeptide fraction from rat liver cell conditioned medium with multiplication-stimulating activity for embryo fibroblasts. *J. Cell. Physiol. 81*, 153-160.

18. Dulbecco, R. (1970). Topoinhibition and serum requirement of transformed and untransformed cells. *Nature 227*, 802-806.

19. Dulbecco, R., and Stoker, M. G. P. (1970). Conditions determining initiation of DNA synthesis in 3T3 cells. *Proc. Nat. Acad. Sci. U.S. 66*, 204-210.

20. Eagle, H. (1965). Metabolic controls in cultured mammalian cells. *Science 148*, 42-51.

21. Flaxman, B. A., Lutzner, M. A., and Van Scott, E. J. (1967). Cell maturation and tissue organization in epithelial outgrowths from skin and buccal mucosa *in vitro. J. Invest. Dermatol. 49*, 322-332.

22. Garcia-Giralt, E., Berumen, L., and Macieira-Coelho, A., (1970). Growth inhibitory activity in the supernatants of nondividing WI-38 cells. *J. Nat. Cancer Inst. 45*, 649-655.

23. Griffiths, J. B. (1972). The effect of cell population density on nutrient uptake and cell metabolism: A comparative study of human diploid and heteroploid cell lines. *J. Cell Sci. 10*, 515-524.

24. Holley, R. W. (1972). A unifying hypothesis concerning the nature of malignant growth. *Proc. Nat. Acad. Sci. U.S. 69*, 2840-2841.

25. Holley, R. W., and Kiernan, J. A. (1968). "Contact inhibition" of cell division in 3T3 cells. *Proc. Nat. Acad. Sci. U.S. 60*, 300-304.

26. Ingram, V. M. (1969). A side view of moving fibroblasts. *Nature 222*, 641-644.

27. Kruse, P. F., Jr., Whittle, W., and Miedema, E. (1969). Mitotic and nonmitotic multiple-layered perfusion cultures. *J. Cell Biol. 42* 113-121.

28. Macieira-Coelho, A. (1967). Dissociation between inhibition of movement and inhibition of division in RSV transformed human fibroblasts. *Exp. Cell Res. 47*, 193-200.

29. Maroudas, N. G. (1972). Anchorage dependence: Correlation between amount of growth and diameter of bead, for single cells grown on individual glass beads. *Exp. Cell Res. 74*, 337-342.

30. Maroudas, N. G. (1973). Growth of fibroblasts on linear and planar anchorages of limiting dimensions. *Exp. Cell Res. 81*, 104-110.

31. Martz, E., and Steinberg, M. S. (1972). The role of cell-cell contact in "contact" inhibition of cell division: A review and new evidence. *J. Cell Physiol. 79*, 189-210.

32. Paul, D., Lipton, A., and Klinger, I. (1971). Serum factor requirements of normal and simian virus 40-transformed 3T3 mouse fibroblasts. *Proc. Nat. Acad. Sci. U.S. 68*, 645-648.

33. Rein, A., and Rubin, H. (1971). On the survival of chick embryo cells at low concentrations in culture. *Exp. Cell Res. 65*, 209-214.

34. Rubin. H. (1971). pH and population density in the regulation of animal cell multiplicaiton. *J. Cell Biol. 51*, 686-702.

35. Rubin, H., and Koide, T. (1973). Inhibition of DNA synthesis in chick embryo cultures by deprivation of either serum or zinc. *J. Cell. Biol. 56*, 777-786.

36. Scher, C. D., and Nelson-Rees, W. A. (1971). Direct isolation and characterization of "flat" SV40-transformed cells. *Nature New Biol. 233*, 263-265.

37. Shodell, M. (1972). Environmental stimuli in the progression of BHK/21 cells through the cell cycle. *Proc. Nat. Acad. Sci. U.S. 69*, 1455-1459.

38. Stoker, M., O'Neill, C., Berryman, S., and Waxman, V. (1968). Anchorage and growth regulation in normal and virus-transformed cells. *Int. J. Cancer 3*, 683-693.

39. Stoker, M. G. P. and Rubin, H. (1967). Density dependent inhibition of cell growth in culture. *Nature 215*, 171-172.

40. Temin, H. M., Pierson, R. W., Jr., and Dulak, N. C. (1972). The role of serum in the control of multiplication of avian and mammalian cells in culture. *In* "Growth, Nutrition and Metabolism of Cells in Culture," Rothblat, G. H. and V. J. Cristofalo, eds. Vol. 1, pp. 49-81. Academic Press, New York.

41. Todaro, G. J., and Green, H. (1963). Quantitative studies of the growth of mouse embryo cells in culture and their development into established cell lines. *J. Cell Biol. 17*, 299-312.

42. Todaro, G. J., Green, H., and Goldberg, B. (1964). Transformation of properties of an established cell line by SV40 and polyoma virus. *Proc. Nat. Acad. Sci. U.S. 51*,66-73.

43. Todaro, G. J., Lazar, G. K., and Green, H. (1965). The initiation of cell division in a contact-inhibited mammalian cell line. *J. Cell. Comp. Physiol. 66*, 325-333.

44. Todaro, G. J., Matsuya, Y., Bloom, S., Robbins, A., and Green, H. (1967). Stimulation of RNA synthesis and cell division in resting cells by a factor present in serum. *In* "Growth Regulating Substances for Animal Cells in Culture," pp. 87-98 (V. Defendi and M. Stoker, eds.) The Wistar Institute Press, Philadelphia, Pennsylvania.

45. Vesely, P. (1972). Tumour cell surface specialization in the uptake of nutrients evidenced by cinemicrography as a phenotypic condition for density independent growth. *Folia Biologica* (Praha) *18*, 395-401.

46. Vesely, P., and Boyde, A. (1973). The significance of SEM evaluation of the cell surface for tumor cell biology. *In*: Scanning Electron Microscopy (Part III), Proc. Workshop on Scanning Electron Microscopy in Pathology, IIT Res. Inst., Chicago, pp. 689-696.

47. Vesely, P., and Weiss, R. A. (1973). Cell locomotion and contact inhibition of normal and neoplastic rat cells. *Int. J. Cancer 11*, 64-76.

48. Westermark, N. F., Brunk, U., Ericsson, J., and Ponten, J. (1972). Studies on in vitro cultivated human glia and glioma cells, I and II. *Acta. Path. Microbiol. Scand. 80A*, 695-696.

49. Zetterberg, A., and Auer, G. (1970). Proliferative activity and cytochemical properties of nuclear chromatin related to local cell density of epithelial cells. *Exp. Cell Res. 62* 262-270.

THE CELL CYCLE AND THE NEOPLASTIC TRANSFORMATION

Jung-Chung Lin and Renato Baserga

Department of Pathology and Fels Research Institute
Temple University School of Medicine
Philadelphia, Pennsylvania 19140

Gross and microscopic differences between normal and neoplastic cells have been described in text books of pathology for over 100 years, and they still constitute the basis for the diagnosis of malignancy in man. Although in an occasional instance the histological diagnosis of malignancy may present difficulties, in general the gross and microscopic changes (invasive growth, metastases, polymorphism of cells, abnormal mitosis, etc.) are so striking that a diagnosis of neoplastic disease can be made with a considerable degree of certainty. In fact, when a pathologist looks at a tumor, the differences between a neoplastic growth and its tissue of origin are of such magnitude that one cannot help thinking that fundamental biochemical differences must be at the basis of the microscopic and gross changes. Biochemical differences between tumors and normal tissues have indeed been described for over 50 years, but despite the thorough efforts of many investigators and some extravagant claims, most biochemical differences have not resisted the onslaught of time. One by one, the biochemical differences between tumors and normal tissues reported in the literature, have turned out to be of limited significance, either because the biochemical alteration was peculiar to only a few special tumors, or because it could be found also in some normal tissues.

Such a preamble ought to discourage any further attempt to describe biochemical differences between neoplastic and normal cells. However, in recent years, more meaningful comparisons have become feasible with the advent of minimal deviation hepatomas and of cell cultures. Differences between liver and minimal deviation hepatomas have been considered in several reviews (Morris, 1963; Morris, 1965), including two recent reviews by Weinhouse *et al.*, 1972; and by Schapira and Hatzfeld, 1972, on certain isozymes whose fetal (absent in the adult liver) reappear in some minimal

This work was supported by U.S.P.H.S. research grants CA-08373 and PO 1-CA-12923 from the National Cancer Institute.

deviation hepatomas. As to cell cultures, even a simple enumeration of all the differences between normal and transformed or neoplastic cells in culture would require a review of several pages. The present discussion will therefore be limited to a comparison of some of the differences between neoplastic and normal cells in culture in relation to the cell cycle, with special emphasis on non-histone chromosomal proteins and their role in the regulation of cell proliferation.

The reader is supposed to be familiar with the cell cycle both from a descriptive and a biochemical aspect. Extensive reviews of the methodology for studying the cell cycle and the kinetics of cellular proliferation have been published by Quastler and Sherman, 1959; Lamerton and Fry, 1963; Feinendegen, 1967; Baserga and Wiebel, 1969; Baserga and Malamud, 1969; Lipkin, 1971; and Gavosto and Pileri, 1971. Several reviews have also been written on the biochemical events occurring during the cell cycle and the reader is referred to a recent book for detailed information (Baserga, 1971, passim). Briefly, it can be said that after mitosis the cell goes through a series of biochemical events, which include the synthesis of special RNA and protein molecules, eventually leading to the onset of DNA synthesis. DNA and histones are replicated during a discrete period of the cell cycle called the S phase, after which another series of steps, which include protein and RNA synthesis, but not DNA synthesis, called the G_0 phase, leads again to mitosis. Nondividing cells leave the cell cycle usually in G_1, sometimes in G_2 and almost never in S or mitosis (Frankfurt, 1967; Epifanova and Terskikh, 1969; and Post et al., 1973). G_0 cells can be distinguished from nondividing cells by their ability to synthesize DNA and divide again when an appropriate stimulus is given. As it will be discussed below, the ability of cells to enter the G_0 state may be an important difference between normal and neoplastic cells.

Contrary to expectation, the length of the cell cycle is not a distinguishing feature between normal and neoplastic cells. It was first reported in 1962 by Baserga and Kisieleski that the cell cycle of continuously dividing normal cells of mice can be shorter than the cell cycle of some of the fastest growing mouse tumors. The authors compared the length of the cell cycle of the lining epithelium of the crypts of the small intestine with that of Ehrlich ascites cells growing in the peritoneal cavity of mice. They pointed out that the growth of tumors in vivo does not necessarily depend on an increased speed of cellular proliferation (that is, on a shortening of the cell cycle), but that it may depend on other factors. Their results have been confirmed by numerous data subsequently gathered in experimental animals, in man, and finally, in vitro by Norrby, 1970. The growth fraction and the rate of cell loss are now generally considered more important than the length of the cell cycle in determining the growth of normal or abnormal tissues (Lipkin, 1971; Gavosot and Pileri, 1971).

The fact that some normal cells may actually proliferate faster than rapidly growing tumors, obviously has serious implications in terms of a cancer chemotherapy based on the use of antimetabolites of nucleic acids. A way out from this impasse could be found in any one of three approaches, namely (1) immunotherapy of tumors; (2) the judicious use of the cell cycle in order to exploit differences between normal and tumor cells; and (3) an understanding of the molecular differences between proliferating normal cells and neoplastic cells. It is this last approach that will occupy us for the rest of the present discussion.

Molecular differences between normal and transformed cells in culture have been described in several instances. In particular, changes in the function and structure of surface membranes in transformed cells have been known for a long time and since they have been reviewed recently by Pardee (1971) and by Burger (1971), will not be examined in detail here. Unfortunately, many of these studies have been vitiated by the fact that investigators have taken as normal cells 3T3 mouse fibroblasts that are disqualified as representatives of normal cells in culture by their complement of chromosomes (near-tetraploid). More interesting are differences described between normal diploid fibroblasts and transformed cells in culture. Thus, Kelly and Sambrook (1973) and Wright and Hayflick (1972), have reported differences in the sensitivity to the drug between normal and transformed fibroblasts in culture. Wright and Hayflick (1972) obtained their results with WI-38 human diploid fibroblasts and their SV-40 transformed counterparts, 2RA cells. Rovera and Baserga (1973) have compared changes in chromatin template activity occurring in WI-38 cells and 3T6 mouse fibroblasts stimulated to proliferate, and Studzinski and Gierthy (1973) have reported a difference in sensitivity to the aminonucleoside of puromycin between WI-38 fibroblasts and HeLa cells. However, the focus of the present discussion will be on the differences in non-histone chromosomal proteins between WI-38 and their SV-40 transformed counterparts, the 2RA cells.

Non-histone chromosomal proteins have recently been implicated as regulators of gene expression in general, and cell proliferation in particular. Again we must refer the reader to previous reviews for the evidence indicating that non-histone chromosomal proteins are possible gene regulators in mammalion cells (See Stellwagen and Cole, 1969; Stein and Baserga, 1972; and McClure and Hnilica, 1972). In this discussion we are mostly concerned with their role in the control of cell proliferation in mammalian cells. Teng and Hamilton (1969) have shown an increased synthesis of non-histone chromosomal proteins in rat uterus stimulated by estrogen, a finding confirmed by Mayol and Thayer (1970), and by Barnea and Gorsky (1970). Baserga and Stein (1971) have reported similar findings in salivary glands of mice stimulated to proliferate by isoproterenol. *In vitro*, a marked stimulation of non-histone chromosomal proteins has been reported in explanted rat mammary glands (Stellwagen and Cole, 1969b) and in WI-38 human diploid fibroblasts stimulated to proliferate by a change of medium (Rovera and Baserga, 1971). This increase in the synthesis of non-histone chromosomal proteins that occurs very early in the prereplicative phase of G_0 cells stimulated to proliferate, is not due to a generalized increase in the synthesis of all classes of non-histone chromosomal proteins. On the contrary, a selective increase in the synthesis of certain non-histone chromosomal proteins in quiescent cells stimulated to proliferate has been described in WI-38 human diploid fibroblasts by Tsuboi and Baserga (1972), in hamster fibroblasts by Becker and Stanners (1972) and in phytohemagglutinin-stimulated lymphocytes by Levy et al. (1973). In all these models of stimulated DNA synthesis the increase in the synthesis of non-histone chromosomal proteins occurred in the first 3 hours or so after stimulation, that is, several hours before the onset of DNA synthesis.

The synthesis of non-histone chromosomal proteins in stationary transformed cells stimulated to proliferate has been described by Tsuboi and Baserga (1972) in 3T6 mouse fibroblasts, but a more meaningful comparison with WI-38 cells could be offered by a study of the synthesis of non-histone chromosomal proteins and other nuclear proteins in SV-40 transformed WI-38, a cell line that goes under the name of 2RA. The rest of the paper will deal with our findings in stationary cultures of 2RA cells stimulated to proliferate by nutritional changes.

MATERIALS AND METHODS

Cell Cultures. SV-40 transformed human diploid fibroblasts (2RA), a gift from Dr. Vincent Cristofalo, Wistar Institute, were used in the present study. The cells were maintained in Falcon plastic T-flasks (surface of 75 sq. cm.) in Basal Medium Eagle (Flow Laboratories) plus 5% fetal calf serum, 2 x vitamins and penicillin (200 units/ml) and streptomycin (94 μg/ml), in an atmosphere of 10% CO_2 and 90% air at 37°C. The cells grown in this medium (BME) reached confluency 7 days after plating.

Stimulation and labeling of proteins. For stimulation experiments, cells were grown for 9 days in L-Isoleucine deficient (2.6 mg/liter) Dulbecco's modified Eagle medium (Grand Island Biological Laboratories, Grand Island, New York) as described by Ley and Tobey (1970). The quiescent cultures were then stimulated with fresh regular MEM (52.4 mg/liter of isoleucine) plus 5% fetal calf serum and 2 x vitamins. At 3-hour intervals, 10 flasks were taken and divided into two groups. The medium was discarded, the cells were washed twice with warm Hanks' balanced salt solution (Micro-biological

Associates, Inc., Bethesda, Maryland), and pulse-labeled with either H^3-leucine (2 μCi/ml) or H^3-tryptophan (2 μCi/ml) in warm Hanks' for 30 minutes. H^3-leucine (specific activity 30.8 Ci/mmole) and H^3-tryptophan (specific activity 2.5 Ci/mmole) were purchased from New England Nuclear Co.

Autoradiography. Cells were grown on cover-slips in small Petri dishes (35 x 10 mm) in L-Isoleucine deficient MEM for 9 days as described above and then stimulated with fresh, complete MEM containing H^3-thymidine (0.05 μCi/m1) supplemented with 5% serum. Cultures were fixed in Carnoy's fixative and autoradiographs were prepared as described by Baserga and Malamud (1969). The emulsion used was Eastman Kodak NTB. The percentage of labeled cells and mitosis were determined on a total of 3000 cells in three different coverslips for each point.

Harvest of the cells. After pulse labeling the cells were washed three times with Ca^{++} - Mg^{++} free Hanks' solution and scraped with a rubber policeman in the same solution. Aliquotes of cell suspensions were taken to determine the incorporation of radioactive aminoacids into soluble and acid insoluble total cellular fractions. The cells were then harvested by centrifugation at 2500 rpm for 3 minutes in an International centrifuge (PR6).

Isolation of Chromatin. (Marushige and Bonner, 1966). Nuclei were isolated by the method of Hymer and Kuff (1964). The cell pellets obtained after centrifugation were vortexed with 8 volumes of 1% Triton X-100, 20 mM EDTA, 80 mM NaCl, pH 7.4. To this, were added 3 volumes of 1 M sucrose in 20 mM EDTA, 80 mM NaCl, pH 7.4. This step was repeated 3 or 4 times until clean nuclei were obtained without cytoplasmic contamination. The clean nuclei were washed twice with 5 ml of cold 0.15 M NaCl, 0.01 M Tris, pH 8.3 and allowed to swell and lyse in 3 ml of cold distilled water for 30 minutes. The nuclei were then gently homogenized with about 30 strokes of a Potter-Elvejhem homogenizer and a uniform chromatin suspension was obtained. The chromatin suspension was layered onto 1.7 M Sucrose, stirred into the upper 2/3 of the tube, and spun in an SV 50.1 Spinco rotor for 80 minutes at 37,000 rpm.

Dissociation of Chromatin. The chromatin obtained after centrifugation through sucrose was resuspended by homogenization in 2.5 ml or 0.1 M sodium phosphate buffer, pH 7.0. Aliquots (0.1 ml) were taken for Oxymat combustion and counted in a Packard Tri-Carb liquid scintillation spectrometer. For dissociation of proteins, chromatin suspensions were adjusted to 10 ml containing 6 M urea, 0.4 M guanidine hydrochloride, 0.1% β-mercaptoethanol, 0.1 M sodium phosphate (ph 7.0), by addition of solid urea, and β-mercaptoethanol (Levy et al., 1972). DNA was sedimented by centrifugation at 35,000 rpm at 4°C for 40 hours, using a 50 Ti rotor in a Spinco L2-65B ultracentrifuge. The supernatant contained the nuclear proteins. DNA pellets were dissolved in 1 M NaOH and incubated at 37°C for 30 minutes. Aliquots of DNA and supernatant were taken for Oxymat combustion and counted. Approximately 75% of all nuclear proteins were dissociated from DNA under these conditions.

Determination of DNA synthesis. Cells were grown in L-Isoleucine deficient MEM for 9 days and stimulated under the same conditions as described above. At various intervals after stimulation, the medium was discarded and the cells were pulse-labeled with H^3-thymidine (0.5 μCi/ml, specific activity 6.7 Ci/mmole) in Hanks' balanced salt solution for 30 minutes. Incorporation of H^3-thymidine into DNA was determined according to the method of Scott *et al.* (1956).

RESULTS AND DISCUSSION

Stimulation of Cell Proliferation in Confluent Monolayers of 2RA cells.

2RA cells do not form a regular monolayer like their WI-38 counterpart, but if the medium is not renewed, they reach, within a few days, a stationary phase. Several methods were tried in order to obtain the most satisfactory stationary phase from which 2RA cells could be maximally stimulated to proliferate again. Empirically it was found that the best conditions for these experiments were obtained when 2RA cells were grown for 9 days in isoleucine-deficient MEM containing 5% fetal calf serum (see above). Under these conditions the cells showed a reduced ability to incorporate thymidine-^{3}H into DNA as determined either biochemically (Fig. 1) or by autoradiography (Fig. 2). If, on day 9, the isoleucine-deficient medium was replaced with fresh, complete MEM containing 5% fetal calf serum, the quiescent 2RA began to proliferate again. The incorporation of thymidine-^{3}H increased promptly but reached a peak at 9 hours after stimulation when it was about 20-fold above control values (Fig. 1). Incorporation of thymidine-^{3}H into the acid soluble fraction of the cell increased within 1 hour to a value of about twice that of control values and then increased again between 6 and 9 hours, but these increases are not of the same magnitude as the marked increase in the incorporation of thymidine-^{3}H into DNA. This was confirmed by autoradiographic studies which indicated that there is, in fact, an increase in the number of DNA synthesizing cells with time after stimulation (Fig. 2). A plateau was reached between 9 and 12 hours, although the maximum number of DNA synthesizing cells (70%) was found at 32 hours. It should be noted that in Fig. 1 the cells were pulse-labeled with thymidine-^{3}H for 30 minutes, whereas in Fig. 2 the cells were continuously labeled with thymidine-^{3}H from the time of stimulation.

Incorporation of Leucine-^{3}H into Cellular and Nuclear Proteins.

Stationary phase 2RA cells were stimulated, as described above, and at various intervals after stimulation they were labeled for 30 minutes with leucine-^{3}H, as described in Methods and Materials. Both the uptake of

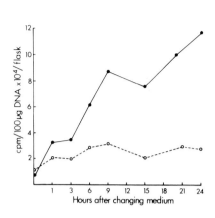

Fig. 1. *Effect of nutritional changes on the incorporation of thymidine-³H into DNA of quiescent 2RA cells. The cells were grown for 9 days in isoleucine-deficient MEM medium. They were then stimulated at 0 time by complete fresh MEM + 5% fetal calf serum and 2x vitamins. At the intervals indicated on the abscissa the medium was discarded and the cultures were pulse-labeled with thymidine-³H (0.5 μCi/ml) in Hanks balanced salt solution for 30 minutes. The uptake of thymidine-³H into the acid soluble fraction and its incorporation into DNA were determined as described in Methods and Materials.* ○———○, *radioactivity in the acid soluble fraction;* ●———●, *radioactivity in DNA. Each point is the average of two determinations from individual Falcon flasks.*

leucine-³H into the acid soluble fraction and its incorporation into total cellular proteins, increased within 1 hour after stimulation. The uptake then remained at this slightly increased value above controls, but the incorporation into total cellular proteins continued to increase and reached a maximum 15 hours after stimulation (Fig. 3). The incorporation of leucine-³H into chromatin and into dissociated chromosomal proteins followed essentially the same pattern (Fig. 4). The maximum was reached between 9 and 12 hours, that is somewhat earlier than the maximum incorporation into total cellular proteins. The increase above control values at 9 and 12 hours was roughly the same for both total cellular proteins and nuclear proteins.

Incorporation of Tryptophan-³H into Total Cellular and Chromosomal Proteins.

Quiescent cultures of 2RA cells were stimulated as described above, and pulse-labeled with tryptophan-³H at various intervals after stimulation. The uptake of tryptophan-³H decreased in the first 3 hours after stimulation and

Fig. 2. *Percentage of cells labeled by thymidine-³H in quiescent cultures of 2RA cells stimulated to proliferate by a nutritional change. Experimental conditions were as shown in Fig. 1, except that the cells were grown on glass cover slips and labeled with thymidine-³H from time 0. After labeling, the cover slips were fixed and processed for autoradiography as described in Methods and Materials.* ●———●, *percent of labeled cells;* ○———○, *percent of mitosis. Each point is the average of determinations from three cover slips.*

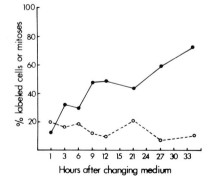

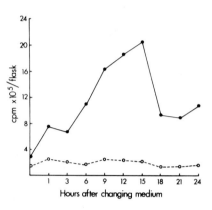

Fig. 3. *Uptake of leucine-³H into the acid soluble fraction and its incorporation into proteins in quiescent 2RA cells stimulated to proliferate. The cells were grown and stimulated as described in Fig. 1. At the intervals indicated on the abscissa the medium was discarded and the cells were labeled with ³H-leucine (2 μCi/ml) in Hanks balanced salt solution for 30 minutes. ○———○, radioactivity in the acid-soluble fraction; ●———●, radioactivity in the acid-insoluble fraction.*

The amount of radioactivity was determined as described in Methods and Materials. Each point is the average of two determinations and each determination is from 5 pooled T-Falcon flasks.

then it rapidly increased, reaching a maximum at about 18 to 21 hours. The incorporation of tryptophan-³H into total cellular proteins instead increased sharply within 1 hour after stimulation, reaching a peak at 12 hours (Fig. 5), 3 hours earlier than the peak obtained with leucine-³H. Incorporation of tryptophan-³H into chromatin and into non-histone chromosomal proteins increased rapidly within 1 hour to values about 5 or 6-fold above control values, vastly exceeding the increase in the incorporation of tryptophan-³H

Fig. 4. *Incorporation of leucine-³H into chromatin and into dissociated chromosomal proteins of quiescent 2RA cells stimulated to proliferate. The cells were grown in isoleucine deficient MEM for 9 days and then stimulated by a change to complete MEM with 5% fetal calf serum and 2x vitamins. At the intervals indicated on the abscissa the medium was discarded and the cells were labeled with leucine-³H (2.0 μCi/ml) in Hanks balanced salt solution for 30 minutes. Chromatin and chromosomal proteins were obtained as described in Methods and Materials. ●———●, incorporation of leucine-³H into chromatin; ○———○, incorporation of leucine-³H into chromosomal proteins.*

Each point is the average of two determinations and each determination is from 5 pooled Falcon flasks.

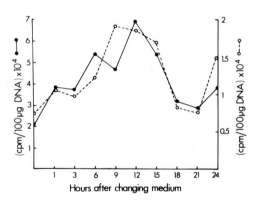

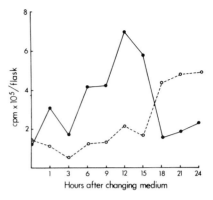

Fig. 5. *Uptake of tryptophan-³H and its incorporation into total cellular proteins of quiescent 2RA cells stimulated to proliferate. Experimental conditions were exactly the same as in Fig. 3, except that leucine-³H was substituted by tryptophan-³H (2.0 μCi/ml).*

into total cellular proteins. The peak incorporation of tryptophan-³H into chromatin and non-histone chromosomal proteins was at 6 hours after stimulation when it was roughly 6 times above control values (Fig. 6). At the same time the incorporation of tryptophan-³H into total cellular protein was 2.5-fold above the control values. These results, therefore, showed that the synthesis of non-histone chromosomal proteins is stimulated in the first 6 hours after quiescent stationary 2RA cells are stimulated to proliferate by nutritional changes.

Other Changes Occurring in Stationary Phase 2RA Cells Stimulated to Proliferate

Costlow and Baserga (1973) have shown that in stationary 2RA cells stimulated to proliferate there is a prompt increase in membrane transport function as evidenced by the ability to take up non-metabolizable amino acids such as cycloleucine. They also showed that after stimulation there is no increase in chromatin template activity at variance with the results obtained in WI-38. The observations of Costlow and Baserga (1973), as well as those of Rovera and Baserga (1973) in WI-38 and 3T6 cells stimulated to proliferate,

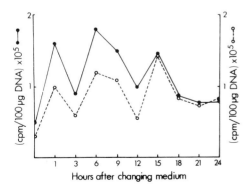

Fig. 6. *Incorporation of tryptophan-³H into chromatin and into dissociated chromosomal proteins of quiescent 2RA cells stimulated to proliferate. Experimental conditions were exactly the same as in Fig. 4, except that leucine-³H was replaced by tryptophan-³H (2.0 μCi/ml).*

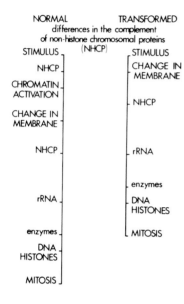

NORMAL   TRANSFORMED
differences in the complement
of non-histone chromosomal proteins
(NHCP)

Fig. 7. *Schematic diagram of the biochemical events occurring in confluent monolayers of normal and transformed cells stimulated to proliferate by nutritional changes. Details are given in the text.*

together with the findings reported in the present discussion, suggest that there are some fundamental differences in the prereplicative phase of WI-38 and 2RA cells stimulated to proliferate by nutritional changes. This difference is schematically outlined in Fig. 7. When confluent monolayers of WI-38 human dipolid fibroblasts are stimulated to proliferate by a change of medium, there is, within 30 minutes after stimulation, an increase in the rate of synthesis of non-histone chromosomal proteins (Rovera and Baserga, 1971). This is followed one hour after stimulation by an increase in the template activity of the isolated chromatin as measured by its ability to incorporate precursors into RNA, using an exogenous *E. coli* RNA polymerase (Farber *et al.*, 1971). At 3 hours after stimulation there is an increase in the ability of the stimulated cells to take up non-metabolizable amino acids such as cycloleucine, (Costlow and Baserga, 1973). This is followed by a second round of increase in the synthesis of non-histone chromosomal proteins (Rovera and Baserga, 1971). Interestingly enough the first round of non-histone chromosomal proteins and the increase in chromatin template activity are insensitive to actinomycin D, whereas all other steps are actinomycin D sensitive. At 8 hours after stimulation there is a peak in the synthesis of ribosomal RNA (Zardi and Baserga, 1974), which is then followed, between 12 and 15 hours, by the replication of DNA and histones and, finally, mitosis. Figure 7 shows that the same sequence of events occurs in stationary 2RA cells stimulated to proliferate except that the first two actinomycin D insensitive steps are not present. Instead, the application of the stimulus causes a prompt increase in the membrane transport function of the 2RA cells, followed in an orderly fashion by the other biochemical events described for the WI-38 cells. Baserga *et al.* (1973), in agreement with similar results obtained by Sander and Pardee (1973) with CHO cells have suggested that the first events occurring in stimulated WI-38 human diploid fibroblasts are part of the G_0 period, whereas all other events are part of the G_1 period. The question naturally arises of why WI-38 are capable of entering G_0, whereas 2RA are incapable, or at least, less capable of doing so. Immunological studies in our laboratory (Zardi *et al.* 1974) have shown that there are substantial differences in the non-histone chromosomal protein complement of WI-38 and 2RA cells. It is postulated that the absence or marked decrease of certain non-histone chromosomal proteins in 2RA cells causes these cells to remain in a G_1 phase after they reach a stationary phase. Since non-histone chromosomal proteins have been implicated in the control of gene expression (see reviews by Stellwagen and Cole, 1969a; and Stein and Baserga, 1972), as well as the control of cell proliferation (see above), it is reasonable to assume that differences in the complement of such proteins may be at the basis of the fundamental differences that exist between these two kinds of cells when they reach a stationary phase. Further studies should elucidate whether these changes can be pinpointed to certain specific non-histone chromosomal proteins as those responsible for the difference in growth characteristics between WI-38 human diploid fibroblasts and their SV-40 transformed counterparts 2RA.

REFERENCES

1. Barnea, A., and Gorski, J. (1970). Estrogen-induced protein: time course of synthesis. *Biochemistry 9*, 1899-1904.

2. Baserga, R. (1971). The Cell Cycle and Cancer. Marcel Dekker, New York.

3. Baserga, R., Costlow, M., and Rovera, G. (1973). Changes in membrane function and chromatin template activity in diploid and transformed cells in culture. *Fed. Proc. 32*, 215-218.

4. Baserga, R., and Kisieleski, W. E. (1962). Comparative study of the kinetics of cellular proliferation of normal and tumorous tissues with the use of tritiated thymidine. I. Dilution of the label and migration of labeled cells. *J. Nat. Cancer Inst. 28*, 331-339.

5. Baserga, R., and Malamud, D. (1969). *In* "Autoradiography: Techniques and Application", p. 253, Harper & Row, Publishers, New York.

6. Baserga, R., and Stein, G. (1971). Nucleic acidic proteins and cell proliferation. *Fed. Proc. 30*, 1752-1759.

7. Baserga, R., and Wiebel, F. (1969). The cell cycle of mammalian cells. *Int. Rev. Exp. Path. 7*, 1-30.

8. Becker, H., and Stanners, C. P. (1972). Control of macromolecular synthesis in proliferating and resting Syrian hamster cells in monolayer culture. III. Electrophoretic patterns of newly synthesized proteins in synchronized proliferating cells and resting cells. *J. Cell. Physiol. 80*, 51-61.

9. Burger, M. M. (1971). Surface changes detected by lectins and implications for growth regulation in normal and in transformed cells. *Biomembranes 2*, 247-270.

10. Costlow, M., and Baserga, R. (1973). Changes in membrane transport function in G_0 and G_1 cells. *J. Cell Physiol. 82*, 411-420.

11. Epifanova, O. I. and Terskikh, V. V. (1969). On the resting periods in the cell life cycle. *Cell Tissue Kinet. 2*, 75-93.

12. Farber, J., Rovera, G. and Baserga, R. (1971). Template activity of chromatin during stimulation of cellular proliferation in human diploid fibroblasts. *Biochem. J. 122*, 189-195.

13. Feinendegen, L. E. (1967). *In* "Tritium-labeled Molecules in Biology and Medicine", Academic Press, New York.

14. Frankfurt, O. S. (1967). Mitotic cycle and cell differentiation in squamous cell carcinomas. *Int. J. Cancer 2*, 304-310.

15. Gavosto, F., and Pileri, A. (1971). Cell Cycle of Cancer Cells in Man. *In* "The Cell Cycle and Cancer", Vol. 1 (R. Baserga, ed.), pp. 99-128, Marcel Dekker, Inc., New York.

16. Hymer, W. C., and Kuff, E. L. (1964). Isolation of nuclei from mammalian tissues through the use of Triton X-100. *J. Histochem. Cytochem. 12*, 359-363.

17. Kelly, F. and Sambrook, J. (1973). Differential effect of cytochalasin B on normal and transformed mouse cells. *Nature New Biol. 242*, 217-219.

18. Lamerton, L. F., and Fry, R. J. M. (1963). "Cell Proliferation", Blackwell, Oxford.

19. Levy, R., Levy, S., Rosenberg, S. A., and Simpson, R. T. (1973). Selective stimulation of non-histone chromatin protein synthesis in lymphoid cells by phytohemagglutinin. *Biochemistry 12*, 224-228.

20. Levy, S., Simpson, R. T., and Sober, H. A. (1972). Fractionation of chromatin components. *Biochemistry 11* (9), 1547-1554.

21. Ley, K. D., and Tobey, R. A. (1970). Regulation of initiation of DNA synthesis in Chinese hamster cells. II. Induction of DNA synthesis and cell division by isoleucine and glutamine in G_1 arrested cells in suspension culture. *J. Cell Biol. 47*, 453-459.

22. Lipkin, M. (1971). The proliferative cycle of mammalian cells. *In* "The Cell Cycle and Cancer", (R. Baserga, ed.), Vol. 1, pp. 6-26, Marcel Dekker, Inc., New York.

23. Marushige, K. and Bonner, J. (1966). Template properties of liver chromatin. *J. Mol. Biol. 15* 160-174.

24. Mayol, R. F., and Thayer, S. A. (1970). Synthesis of estrogen-specific proteins in the uterus of the immature rat. *Biochemistry 9*, 2484-2489.

25. McClure, M. E., and Hnilica, L. S. (1972). Nuclear proteins in genetic restriction. III. The Cell Cycle. *Sub-Cell. Biochem. 1*, 311-332.

26. Morris, H. P. (1963). Some growth, morphological and biochemical characteristics of hepatoma 5123 and other new transplantable hepatomas. *Progr. Exp. Tumor Res. 3*, 370-411.

27. Morris, H. P. (1965). Studies on the development, biochemistry and biology of experimental hepatomas. *Advan. Cancer Res. 9*, 227-302.

28. Norrby, K. (1970). Population kinetics of normal, transforming and neoplastic cell lines. *Acta Path. Microbiol. Scand. 78*, suppl. 214, 1-50.

29. Pardee, A. B. (1971). The surface membrane as a regulator of animal cell division. *In Vitro 7*, 95-104.

30. Post, J., Sklarew, R. J., and Hoffman, J. (1973). Cell proliferation patterns in autogenous rat sarcoma. *J. Nat. Cancer Inst. 50*, 403-414.

31. Quastler, H., and Sherman, F. G. (1959). Cell population kinetics in the intestinal epithelium of the mouse. *Exp. Cell Res. 17*, 420-438.

32. Rovera, G., and Baserga, R. (1971). Early changes in the synthesis of acidic nuclear proteins in human diploid fibroblasts stimulated to synthesize DNA by changing the medium. *J. Cell Physiol. 77*, 201-212.

33. Rovera, G., and Baserga, R. (1973). Effect of nutritional changes on chromatin template activity and non-histone chromosomal protein synthesis in WI-38 and 3T6 cells. *Exp. Cell Res. 78*, 118-126.

34. Sander, G., and Pardee, A. B. (1972). Transport changes in synchronously growing CHO and L Cells. *J. Cell. Physiol. 80*, 267-272.

35. Scott, J. P., Fraccastoro, A. P., and Taft, E. B. (1956). Studies in histochemistry. I. Determination of nucleic acids in microgram amount of tissue. *J. Histochem. Cytochem. 4*, 1-10.

36. Stein, G., and Baserga, R. (1972). Nuclear proteins and the cell cycle. *Adv. Cancer Res. 15*, 287-330.

37. Stellwagen, R. H., and Cole, R. D. (1969). Chromosomal proteins. *Ann. Rev. Biochem. 38*, 951-990.

38. Stellwagen, R. H., and Cole, R. D. (1969). Histone biosynthesis in the mammary gland during development and lactation. *J. Biol. Chem. 244*, 4878-4887.

39. Studzinski, G. P., and Gierthy, J. F. (1973). Selective inhibition of the cell cycle of cultured human diploid fibroblasts by aminonucleoside of puromycin. *J. Cell. Physiol. 81*, 71-84.

40. Teng, C. S., and Hamilton, T. H. (1969). Role of chromatin in estrogen action in the uterus. II. Hormone-induced synthesis of non-histone acidic proteins which restore histone-inhibited DNA-dependent RNA synthesis. *Proc. Nat. Acad. Sci. U.S.A. 63*, 465-472.

41. Tsuboi, A., and Baserga, R. (1972). Synthesis of nuclear acidic proteins in density-inhibited fibroblasts stimulated to proliferate. *J. Cell. Physiol. 80*, 107-118.

42. Weinhouse, S., Shatton, J. B., Criss, W. E., and Morris, H. P. (1972). Molecular forms of Enzymes in Cancer. *Biochimie 54*, 685-693.

43. Wright, W. E., and Hayflick, L. (1972). Formation of anucleate and multinucleate cells in normal and SV_{40} transformed WI-38 by cytochalasin B. *Exp. Cell Res. 74*, 187-194.

44. Zardi, L., Lin, J., Petersen, R. D., and Baserga, R. (1974). Specificity of antibodies to non-histone chromosomal proteins of cultured fibroblasts. *In* "Control of Cell Proliferation in Animal Cells". Cold Spring Harbor Symp. (in press).

45. Zardi, L., and Baserga, R. (1974). Ribosomal RNA Synthesis in WI-38 Cells stimulated to proliferate. Exptl. Molec. Path. (in press).

NEOPLASIA AND DIFFERENTIATION AS TRANSLATIONAL FUNCTIONS[1]

Henry C. Pitot

McArdle Laboratory for Cancer Research
University of Wisconsin School of Medicine
Madison, Wisconsin 53706

The development of a multicolonal organism from a single cell, a characteristic of all multicellular eukaryotic organisms, is a major biological phenomenon, the mechanism of which is as yet not understood. Perhaps the most intriguing aspect of this entire developmental sequence is the production of clones of different but stable phenotypes originating from a single cell, all cells possessing identical genotypes. The critical problem facing the developmental biologist is the elucidation of the mechanisms by which such stable phenotypes develop from a single genotype.

Considerable speculation and experimental effort has been directed towards an elucidation of the mechanism of differentiation. "Inducers" and other gene regulating substances (Waddington, 1967) have long been sought after and described in developmental biology. Theoretical concepts of gene activation involving a system of gene slaves (Callan, 1967; Britten and Davidson, 1970; Davidson and Britten, 1971) have been described to account for epigenetic changes which occur during developmental sequences in eukaryotic organisms. The relationship of cell interactions (Grobstein, 1966) and the importance of surface-associated mucopolysaccharide-protein complexes as well as contractile filamentous organelles in this process (Bernfield and Wessels, 1970) have also been implicated in this process. The peculiar characteristics of cellular differentiation in certain mesenchymal tissues may present unique problems for the mechanistic theoreticians (Hall, 1970). More recently, the organ-specific messenger RNA populations of differentiated tissues (Paul and Gilmour, 1968) have been related to histone and other basic protein populations and their interactions with DNA in the nucleus (Bonner, *et al.*, 1968). In addition, the potential significance of repetitive DNA

[1] Portions of the experimental data reported herein were supported by grants from the National Cancer Institute (CA-07175) and the American Cancer Society (E-588).

sequences and gene duplication in the regulation of the differentiation of cellular populations during the developmental sequence has been emphasized (Ohno, 1972; Britten and Davidson, 1971).

However, despite numerous theories and endless hours in the laboratory, our understanding of the basic mechanisms of cellular differentiation is as yet quite incomplete. On the other hand, the ramifications of these studies to other areas of biology and pathobiology are now beginning to become apparent.

The Neoplastic Transformation as a Problem in Developmental Biology

The possible relationship of cancer to embryogenesis and differentiation is not a new consideration. Connheim (1889) promoted the "embryonic rest theory" of cancer near the turn of the century and quoted a significant amount of biological evidence in support of such a concept. The term "de-differentiation" has been utilized by pathologists for many years in their morphologic descriptions of neoplasms, especially those that are most aggressive. The actual classification of malignant neoplasms is based on their developmental characteristics. In addition, as seen in Table 1, other attributes, both of carcinogenesis and of cancer itself, have their analogues in developmental biology. Obviously, such comparisons do not prove a necessary relationship between these two processes, but this evidence does warrant further study. Because of limited space and time this discussion will be limited to the last point in Table 1, that of messenger RNA stability in neoplasia and during differentiation.

Messenger RNA Template Stabilization in the Regulation of Genetic Expression

Since differentiation is clearly a problem in the regulation of genetic expression, most investigators have concerned themselves with the importance

TABLE 1.

Characteristics of Developmental Biology and Pathobiology of Neoplasia

Developmental	Neoplasia
Viruses induce developmental abnormalities	Viruses induce neoplasia
Organ specific antigens develop during embryogenesis	Tumor specific antigens develop during carcinogenesis
Specific chemicals are teratogenic	Specific chemicals are carcinogenic
Relatively high glycolytic rate in most embryonic tissues	Relatively high glycolytic rate in most neoplastic tissues
mRNA template stabilization during differentiation	Altered mRNA template stability in the several neoplasms studied

and significance of specific sites in the biochemical mechanisms of genetic expression as key points in the regulation of differentiation. Several of these, such as gene duplication and histones, have been referred to above. One of the relatively unique characteristics of eukaryotic cells is their apparent ability to stabilize in the cytoplasm previously synthesized messenger RNA templates (Pitot, 1967). While definitely not unique to the eukaryotic system, this mechanism further requires that the biologic system develop distinct mechanisms for the regulation of genetic translation by environmental means. Several translational or post-translational regulatory events have been described in the adult organism composed of numerous populations, most of which are not actively undergoing cell division but rather are performing specific functions for the organismal economy with their cells predominantly in the resting phase. In such systems protein degradation occurs as does protein synthesis. Several investigations have demonstrated the environmental regulation of protein degradation, a phenomena which one may argue is a consequence of the presence of stable messenger RNAs within the cell. Schimke and his collaborators (1965) have demonstrated that tryptophan administration stabilizes the hepatic enzyme, tryptophan oxygenase, *in vivo* preventing its normal turnover. Cortisone has been shown to totally abolish the turnover of tyrosine aminotransferase for short periods of time after its administration *in vivo* (Levitan and Webb, 1969).

While protein degradation may or may not be closely related to genetic translation, other data have definitely shown that the rate of translation may be regulated by environmental factors. Jost *et al.* (1970) demonstrated that the administration of glucose to rats *in vivo* completely repressed the synthesis of the enzyme, serine dehydratase, shortly after intubation of the carbohydrate. This effect occurred even during periods when synthesis of the enzyme was totally independent of genetic transcription as evidenced by the use of actinomycin D. In addition, Cho-Chung and Pitot (1968) demonstrated that tryptophan administration to animals maintained on a low protein diet not only prevented tryptophan oxygenase degradation but also doubled the rate of synthesis of the enzyme. This effect was totally independent of RNA synthesis. Recent studies in our laboratory by Iwasaki (Iwasaki, *et al*, 1973) have demonstrated the existence of multiple forms of tyrosine aminotransferase. The synthesis of three of these forms are clearly regulated by different environmental mechanisms, one of which involves insulin. Physicochemical measurements have indicated that 3 of the 4 forms are quite similar and probably differ one from another by some chemical modification of the primary gene product (Inoue *et al* 1971). In neonatal animals, the insulin regulated form is normally not present but after priming the animal with cortisone, it is possible to show that the synthesis of this form may be "induced" by insulin largely in the absence of RNA synthesis.

These experiments show rather conclusively that the rate of translation and the rate of protein degradation, a post-translational event, may be regulated by environmental means in adult tissues. They further demonstrate not only the existence of stable messenger RNA templates, but also the environmental regulation of their expression in the cytoplasm of the cell.

Template Stabilization during Embryonic Development and in Normal Adult Tissues

The phenomenon of template stabilization during the developmental sequence has now been documented in many laboratories with a number of eukaryotic systems. In Table 2 are listed a number of examples shown experimentally where messenger RNA templates for specific proteins are stabilized during development, usually at specified times in the normal sequence of cellular differentiation. In most instances the experimental determination of transcriptional or translational regulation is made by means of the inhibition of RNA synthesis with subsequent determination of the effect of this on translation. The use of the antibiotic, actinomycin D, in such experiments is not without certain difficulties as have recently been pointed out. Singer and Penman (1972) demonstrated that in tissue culture the administration of actinomycin tends to decrease the apparent messenger RNA stability, probably through effects on translation itself. On the other hand, Endo et al. (1972) showed that the administration of actinomycin D *in vivo* actually causes a prolongation of the over-all "half life" of liver messenger RNA. These studies and those of Murphy and Attardi (1973) do not in fact negate the data seen in Table 2 since much of that data was carried out *in vitro* suggesting that if anything, the use of the inhibitor should not have demonstrated stabilization but rather inactivation of messenger RNA.

In adult systems, besides that described by Balis and his collaborators (1971) in the small intestine and Bresnick et al. (1967) in regenerating liver, our laboratory has been interested in the question of messenger RNA templates stability and its role in the neoplastic transformation (Pitot, 1968). Some years ago we described the change in the sensitivity of the synthesis of serine dehydratase induction to the administration of actinomycin D *in vivo* (Pitot and Peraino, 1963). These data argued strongly that while the initiation of induction of the enzyme required both transcription and translation within one to two hours after enzyme synthesis was initially induced, further translation become independent of transcription. As noted above, glucose may completely suppress translation and amino acids may actually increase the rate of translation in the absence of transcription in this system. Further studies utilizing timed pulses of inducing agent in the presence and absence of actinomycin D (Pitot et al., 1965) demonstrated that the synthesis of serine dehydratase remained totally translational for 6 to 8 hours, this period being

TABLE 2.

Alteration of mRNA Template Stability During Normal Differentiation

Enzyme	Tissue	Change in mRNA Template Stability with Differentiation	References
Amylase chymotrypsinogen	Murine Pancreas (11-16 days)	increased mRNA stability	(Rutter et al, 1963)
Glutamine Synthetase	Chick Retina (12-18 days)	increased mRNA stability	(Moscona and Kirk, 1965) (Kirk, 1965)
Crystallin	Differentiation of lens epithelium to fiber cell	increased mRNA stability	(Stewart and Papaconstantinou, 1967)
Cocoonase	Silk moth mouth gland differentiation	increased mRNA stability	(Kafatos and Reich, 1968)
Hemoglobin	Chick blood islands (20-30 hrs)	increased mRNA stability	(Wilt, 1965)
Adenosine Deaminase	Differentiation of crypt to villus cell in rat intestine	decreased mRNA stability	(Imondi, et al, 1970)
Thymidine Kinase Aspartate Transcarbamylase	Regenerating Rat Liver	increased mRNA stability	(Bresnick, et al, 1967)

termed the template life-time. These studies were supported by later investigations (Pitot *et al.*, 1971) utilizing the actual rate of translation measured immunochemically with essentially the same results. Using similar techniques it is possible to estimate certain parameters of the template lifetimes of other enzymes in normal liver. These studies have shown that for each of 5 enzymes studied, the template lifetimes are significantly different.

Therefore, unlike protein turnover which appears to be an exponential decay function, dependent on the amount of the protein molecules in the population at any one time, turnover of messenger RNA, especially those with finite long lifetimes, is not a totally random process. Rather our studies argue that for stable mRNA templates, stability is complete for a definite period of time. Thus, if one were to examine the decay of a single stable message, one would expect a linear rather than a logarithmic function. This is diagrammatically shown in Figure 1 where an asynchronous population of stable messenger RNAs for a single gene are represented. The length of the line is the length of the messenger RNA template life. No matter what the asynchrony of the population, the inhibition of RNA synthesis will lead to a linear decay in the number of active templates. Since, however, in all experiments performed to date the mRNA template lifetime is inferred by measuring either the rate of synthesis or the activity of an enzyme whose decay is exponential, it has thus far been difficult, if not impossible to prove the previous argument. In fact while the decay curve of a single species of messenger RNA may be linear when the message itself is measured directly, measurement of its product will lead to an exponential decay complicated by the linear decay of the mRNA, thus actually demonstrating a biphasic or more complex kinetic appearance. On the other hand, if multiple messenger RNA populations are measured, each having its own characteristic lifetime, it is difficult to predict the type of curve which may occur. In this laboratory, the decay of poly(A)-RNA sequences in the cytoplasm of both liver and hepatoma have been determined as a measure of the total turnover of messenger RNA (Tweedie and Pitot, 1974). Some of this data is seen in Figure 2 wherein the data is plotted as a biphasic linear decay with a break occurring at about the 24 hour time point. The data may be plotted as an exponential decay. Thus,

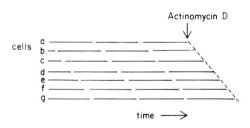

Fig. 1. *Graphic representation of cellular template longevity in asynchronous cells ("a" through "g"). Horizontal lines denote hypothetical template lives of individual cells, the finite life spans being indicated by the length of the line. Blockage of template replacement following treatment with the inhibitor, actinomycin D, is shown by the oblique line.*

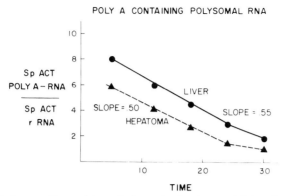

Fig. 2. *Relative specific activity of the polyadenylate-containing fractions of the RNA extracted at pH 9.0 from polyribosomes from rat liver and the Morris hepatoma 7800. There was considerable variation in absolute specific activity of corresponding fractions between rats in this experiment. Thus, the results are expressed as the ratio of the specific activity of the polyadenylate-containing RNA fraction to the specific activity of the RNA fraction extracted at pH 7.6 which is predominantly ribosomal RNA. The time intervals are in hours and the label was $^{32}PO_4$ given at 0 time (1 millicurie per rat).*

unless extremely careful and numerous measurements are made, it may be difficult to distinguish between a highly complex decay curve and that of simple exponential decay.

Altered Template Stability in Neoplasms

Utilizing the same techniques described above, several studies have shown that neoplasms possess mRNA template stabilities which are quite different for their tissue of origin. Hilf and his associates (1967) demonstrated that the induction of several mammary enzymes required both transcription and translation in normal mammary tissue but only required translation in malignant mammary tissue. Recently, Wu *et al.* (1971) demonstrated that actinomycin D administration actually enhanced the cortisol-induced increases

TABLE 3.
Estimated mRNA Template Lifetimes for Several Enzymes in Rat Liver and Hepatomas

Enzyme	Liver	Hepatomas[1]	
		H-35	5123
Serine Dehydratase	6-8 hrs	<1 hr	>2 wks
Ornithine Aminotransferase	18-24 hrs	–	>48 hrs
Tryptophan Oxygenase	>2 wks	<12 hrs	–
Tyrosine Aminotransferase (induced)	2-3 hrs	>6 hrs	–
Thymidine Kinase	<3 hrs	>12 hrs	–

[1] The tumors used were the Reuber H-35 and Morris 5123 hepatocellular carcinomas.

in liver arginase and argininosuccinate synthetase in the Morris 8999 hepatoma while inhibiting the cortisol induction of arginase in the Morris 7800 hepatoma. No effect of the antibiotic was seen in the host liver. In our laboratory we have been able to demonstrate that several enzymes whose characteristics of template stability are known for normal liver exhibit different parameters in neoplastic liver. Some of this data is seen in Table 3 where normal liver and 2 hepatomas are compared. One may see that the template lifetimes of the enzymes in the neoplasms are quite different one from another and also from normal liver for each of the enzymes studied. Like regenerating liver (Bresnick *et al.*, 1967), the template lifetime for thymidine kinase is markedly extended in a hepatoma as compared to host liver tissue. While these studies are obviously relatively few and incomplete, they tend to support the concept that messenger RNA template stabilities are quite abnormal in at least this system of neoplasms.

Conclusions

Several authors have proposed previously that the malignant transformation results from an abnormality in cellular differentiation (Markert, 1968; Pierce, 1970). While template stability is obviously only a single parameter of developmental biology and the neoplastic process, the total significance of changes in template stabilization has yet to be appreciated from the experimental data available. Kafatos (1972) has argued that messenger RNA template lifetimes and turnover may be the major mechanism whereby cellular differentiation is effected. Our own studies (Pitot, 1963) have argued from a theoretical standpoint that the malignant transformation may well result from altered messenger RNA stability, those most malignant tumors having very stable templates of messenger RNAs whose products are directly involved in cellular replication.

Obviously one of the major factors which we have not considered in this discussion is the mechanism of the stabilization of messenger RNA. This laboratory (Pitot, 1969), and more recently others (Kenney *et al.*, 1972) have proposed that messenger RNA stabilization is the result of the interaction of the template with intracellular membrane systems. Such a mechanism could involve a hereditary component similar to that of the external membrane pattern of paramecium (Sonneborn, 1968) while recent studies by Roobol and Rabin (1971) have suggested that hormonal action may affect membrane-polysome interaction and through the theoretical arguments from this laboratory (Pitot, 1968), the stabilization of messenger RNA template. While as yet the evidence is quite circumstantial, future endeavors in this area may well relate neoplasia as an abnormality in differentiation through abnormal template stabilization and translational control.

BIBLIOGRAPHY

1. Bernfield, M. R. and Wessels, N. K. (1970). Intra- and extracellular control of epithelial morphogenesis. Developmental Biology Supplement, *4*, 195-249.

2. Bonner, J., Dahmus, M. E., Fambrough, D., Huang, R. C., Marushige, K. and Tuan, D. Y. H. (1968). The biology of isolated chromatin. Science, *159*, 47-56.

3. Bresnick, E., Williams, S. S. and Mosse, H. (1967). Rates of turnover of deoxythymidine kinase and of its template RNA in regenerating and control liver. Cancer Res. *27*, 469-475.

4. Britten, R. J. and Davidson, E. H. (1969). Gene regulation for higher cells. Science, *165*, 349-357.

5. Britten, R. J. and Davidson, E. H. (1971). Repetitive and non-repetitive DNA sequences and a speculation on the origin of evolutionary novelty. Quarterly Rev. Biol. *46*, 111-138.

6. Callan, H. C. (1967). The organization of genetic units in chromosomes. J. Cell Sci. *2*, 1-7.

7. Cho-Chung, Y. S. and Pitot, H. C. (1968). Regulatory effects of nicotinamide on tryptophan pyrrolase synthesis in rat liver *in vivo*. Eur. J. Biochem. *3*, 401-406.

8. Connheim, J. F. "Ghnheim's Lectures on General Pathology". New Sydenham Society, London, 1889.

9. Davidson, E. H. and Britten, R. J. (1971). Note on the control of gene expression during development. J. Theoretical Biol. *32*, 123-130.

10. Endo, Y., Tominaga, H. and Natori, Y. (1971). Effect of actinomycin D on turnover rate of messenger ribonucleic acid in rat liver. Biochem. Biophys. Acta, *240*, 215-217.

11. Grobstein, C. (1966). What we do not know about differentiation. Amer. Zoologist, *6*, 89-95.

12. Hall, B. K. (1970). Cellular differentiation in skeletal tissues. Biol. Rev. *45*, 455-484.

13. Imondi, A. R., Lipkin, N. and Balis, M. E. (1970). Enzyme and template stability as regulatory mechanisms in differentiating intestinal epithelial cells. J. Biol. Chem. *245*, 2194-2198.

14. Inoue, H., Kasper, C. B. and Pitot, H. C. (1971). Studies on the induction and repression of enzymes in rat liver. VI. Some properties and the metabolic regulation of two isozymic forms of serine dehydratase. J. Biol. Chem. *246*, 2626-2632.

15. Iwasaki, Y., Lamar, C., Danenberg, K. and Pitot, H. C. (1973). Studies on the induction and repression of enzymes in rat liver — characterization and metabolic regulation of multiple forms of tyrosine aminotransferase. Eur. J. Biochem. *34*, 347-357.

16. Yost, J-P., Khairallah, E. A. and Pitot, H. C. (1968). Studies on the induction and repression of enzymes in rat liver. V. Regulation of the rate of synthesis and degradation of serine dehydratase by dietary amino acids and glucose. J. Biol. Chem. *243*, 3057-3066.

17. Kafatos, F. C. (1972). mRNA stability and cellular differentiation. Karolinska Symposia on Research Methods in Reproductive Endocrinology, Vol. 5, pp. 319-341.

18. Kafatos, F. C. and Reich, J. (1968). Stability of differentiation-specific and nonspecific messenger RNA in insect cells. Proc. Natl. Acad. Sci. *60*, 1458-1465.

19. Kenney, F. T., Lee, K-L. and Stiles, C. D. (1972). Degradation of messenger RNA in mammalian cells. Karolinska Symposia on Research Methods in Reproductive Endocrinology, Vol. 5, pp. 369-380.

20. Kirk, D. L. (1965). The role of RNA synthesis in the production of glutamine synthetase by developing chick neural retina. Proc. Nat. Acad. Sci. *54*, 1345-1353.

21. Levitan, I. B. and Webb, T. E. (1970). Hydrocortisone-mediated changes in the concentration of tryosine transaminase in rat liver: an immunochemical study. J. Mol. Biol. *48*, 339-348.
22. Markert, C. L. (1968). Neoplasia: a disease of cell differentiation. Cancer Res. *28*, 1908-1914.
23. Moscona, A. A. and Kirk, D. L. (1965). Control of glutamine synthetase in the embryonic retina *in vitro*. Science, *148*, 519-521.
24. Murphy, W. and Attardi, G. (1973). Stability of cytoplasmic messenger RNA in HeLa cells. Proc. Nat. Acad. Sci. *70*, 115-119.
25. Ohno, S. (1972). Gene duplication, mutation load and mammalian genetic regulatory systems. J. Med. Genet. *9*, 254-263.
26. Paul, J. and Gilmour, R. S. (1968). Organ-specific restriction of transcription in mammalian chromatin. J. Mol. Biol. *34*, 305-316.
27. Pierce, G. B. (1970). Differentiation of normal and malignant cells. Fed. Proc. *29*, 1248-1254.
28. Pitot, H. C. (1964). Altered template stability: the molecular mask of malignancy? Persp. Biol. Med. *8*, 50-70.
29. Pitot, H. C. (1967). "Metabolic regulation in metazoan systems". In: *Molecular Genetics, Part 2*. J. H. Taylor, Ed. Academic Press, New York. pp. 383-423.
30. Pitot, H. C. (1969). Endoplasmic reticulum and phenotypic variability in normal and neoplastic liver. Arch. Path. *87*, 212-222.
31. Pitot, H. C., Kaplan, J. and Cihak, A. (1971). Translational regulation of enzyme levels in liver. In: "Enzyme Synthesis and Degradation in Mammalian Systems". M. Rechcigl, Ed. S. Karger, Basel, pp. 216-235.
32. Pitot, H. C. and Peraino, C. (1964). Studies on the induction and repression of enzymes in rat liver. I. Induction of threonine dehydrase and ornithine- - transaminase by oral intubation of casein hydrolysate. J. Biol. Chem. *239*, 1783-1788.
33. Pitot, H. C., Peraino, C., Lamar, C. Jr. and Kennan, A. L. (1965). Template stability of some enzymes in rat liver and hepatoma. Proc. Nat. Acad. Sci. *54*, 845-851.
34. Roobol, A. and Rabin, B. R. (1971). The binding of polysomes to smooth membranes of rat liver promoted by steroid hormones and extracts from either rough endoplasmic reticulum or from polysomes of the opposite sex. FEBS Letters, *14*, 165-169.
35. Rutter, W. J., Wessels, N. K. and Grobstein, C. (1964). Control of specific synthesis in the developing pancreas. Natl. Cancer Inst. Monograph, *13*, 51-62.
36. Schimke, R. T., Sweeney, E. W. and Berlin, C. M. (1965). The roles of synthesis and degradation in the control of rat liver tryptophan pyrrolase. J. Biol. Chem. *240*, 322-331.
37. Singer, R. H. and Penman, S. (1972). Stability of HeLa cell mRNA in actinomycin. Nature, *240*, 100-102.
38. Stewart, J. A. and Papaconstantinou, J. (1967). A stabilization of RNA templates in lens cell differentiation. Proc. Nat. Acad. Sci. *58*, 95-102.
39. Tweedie, J. W. and Pitot, H. C. (1974). Polyadenylate-containing RNA of polyribosomes isolated from rat liver and Morris hepatoma 7800. Cancer Res. *34*, 109-114.
40. Waddington, C. H. (1967). "Principles of development and differentiation". Macmillan, New York.
41. Wilt, F. H. (1965). Regulation of the initiation of chick embryo hemoglobin synthesis. J. Mol. Biol. *12* 331-339.
42. Wu, C., Bauer, J. M. and Morris, H. P. (1971). Responsiveness of two urea cyclic enzymes in Morris hepatomas to metabolic modulations. Cancer Res. *31*, 12-18.

SESSION III
Cell Surface Changes

Moderator: Murray Rosenberg

OBSERVATIONS ON THE MOBILITY OF PROTEIN COMPONENTS OF THE PLASMA MEMBRANE

Michael Edidin

Department of Biology
The Johns Hopkins University
Charles and 34th Streets
Baltimore, Maryland 21218

Motions on many time scales occur in a cell's plasma membrane. Besides the most rapid of these motions, in the nanosecond to microsecond time range, we find movements of molecules and groups of molecules of distances of 10 to 10,000 Å in times ranging from around a millisecond to hundreds of seconds. The most conspicuous of these relatively slow movements are the extensions of whole areas of membrane in ruffles, blebs and microvilli (Abercrombie, 1961; Harris, 1972; Porter *et al.*, 1973), movements in which several microns of linear distance may be covered in some tens of seconds. These movements are driven by cell metabolism and seem to involve a coherent and elastic membrane (Weiss and Clement, 1969).

Beneath such flamboyant extension, a second set of motions, of individual molecules occurs within the plane of the membrane. These motions seem to be, like those of diffusion, random. We suspect that ultimately they will be seen to be as important to cell physiology as membrane extensions are to cell social behaviour. In this essay I hope to discuss the evidence that diffusion of proteins may occur in the plane of the membrane, mainly discussing ways in which this mobility may be demonstrated. It should be evident that mobility, if found, may be manipulated to gain insight into the organization of the cell surface and into the consequences of diffusion of its component molecules. Such manipulation may be found useful in developing systems, for example, in investigating fertilization, or cell adhesion, or response to hormones and tactic factors.

Acknowledgements: Original work described in this paper was supported by a National Institutes of Health training grant to the Department of Biology and by a research grant AM 11202 to M.E.

This is contribution no. 776 from the Department of Biology.

Methods of demonstrating protein movements in the plasma membrane

Three approaches to showing lateral diffusion of membrane proteins have been used in our laboratory, and by others. These are:

1. Mixing membranes bearing distinctive markers with one another, by cell fusion. The rate of intermixing of the markers may be used as a measure of the ease of their lateral movement. If metabolic effects on movement can be ruled out, then some estimate of diffusion constants or membrane viscosity may be made. While in some cases mixing rates and mechanisms were not followed closely we do have information on the mixing of surface enzymes (Gordon and Cohn, 1970), species antigens (Watkins and Grace, 1967; Harris et al., 1969) and transplantation antigens (Frye and Edidin, 1970).

2. Marking portions of an otherwise homogeneous membrane, then following the spread of the mark. Some of the first experiments showing molecular movement in the plane of the membrane were done in this way. Marking has been achieved by visualization of newly synthesized hemagglutinin sites with erythrocytes (Marcus, 1962), by bleaching a portion of the rhodopsin molecules in a disc (Poo and Cone, 1974), by spotting intact antibody onto a part of the surface of ameba (Wolpert and O'Neill, 1962) and by spotting antibody Fab fragments on the surface of cultured myotubes (Edidin and Fambrough, 1973).

3. Perturbing the cell surface by binding a multivalent ligand. A great variety of experiments have been done in which aggregates of membrane antigens and lectin receptors are induced to collect at one portion of a cell. The most thorough analyses of the phenomenon indicate that it can only take place in a cell whose surface proteins are free to move, but that this freedom of movement is not sufficient to cause large aggregate formation or "capping" (Taylor et al., 1971). Capping is a useful method for modifying the cell surface and for analyzing the relationship of one part to another, but has limited value in quantitating rates of motion.

Antigen mixing in virus-induced heterokaryons

Membranes bearing diverse markers may be mixed by fusion of cells, usually with inactivated Sendai virus. It was early shown that stable hybrids resulting from such fusions carried the surface antigens of both parents (Harris et al., 1969; Gershon and Sachs, 1963), and recently other markers, such as acetylcholine receptors have been found to persist on some hybrids. Both types of marker are found uniformly distributed over the entire cell surface. When early products of virus-induced fusion, heterokaryons, were examined within days or even hours of formation, they too were found to bear intermixed surface antigens (Watkins and Grace, 1967; Harris et al., 1969) or enzymes (Gordon and Cohn, 1970) derived from parent cells. While the uniform distribution of membrane markers in hybrids could well be due to turnover and random insertion of membrane proteins, it was not clear that the mixing in heterokaryons could be accounted for in the same way.

Work in our laboratory by Dr. Larry D. Frye (Frye and Edidin, 1970) showed that indeed another mechanism operated to intermix the surface antigens of heterokaryons, diffusion. The system used mouse and human cell parents: Clld, a clone of L cell fibroblasts, and VA-2 an SV40-transformed human fibroblast line. After fusion was induced by virus the cells were incubated for short times – from a few minutes to around two hours, and then the population's surface antigens were visualized with fluorescent antibodies. Mouse H-2 transplantation antigens and human species antigens (proteins or glycoprotein) were first reacted with appropriate specific antibodies and these in turn were visualized by addition of fluorescent antiglobulins, green fluorescein-conjugated goat anti-mouse IgG and red tetramethylrhodamine-conjugated goat anti-rabbit IgG. The two important technical points to be kept in mind here (aside from the care required to make all antibodies specific for one group of antigens only) are: 1) indirect immunofluorescence is a method of approximately 10-fold greater sensitivity than direct immunofluorescence allowing detection of areas of low antigen concentration, and 2) antibodies are added only after the population of newly-formed heterokaryons have been held for whatever time they are to be tested for intermixing. This means that the perturbing effects of antibody itself have been avoided; reagents to visualize the cell surface are added only after the surface has been given a chance to rearrange. While this is an advantage, I must also note a disadvantage. We have never followed mixing of antigens in a single cell; rather, measurements are made on the change in pattern of staining in a population of cells over time.

Frye found that most of the members of a population of heterokaryons, all of which fluoresced both green and red, bore separate areas of green and red fluorescence when the population was examined between 5 and 10 minutes after initiating fusion at 37°C. Such a segregate heterokaryon is shown in Figure 1. It shows clearly areas that fluoresce in one color but not the second, reporting the presence of mouse and human surface antigens in separate parts of the heterokaryon membrane.

When the population of heterokaryons is sampled about an hour after initiating fusion the proportion of heterokaryons in the entire cell population has not increased drastically, but the majority of these cells are found to bear intermixed surface antigens; they were termed "mosaic" cells by Frye. Figure 2 shows such a cell; while the distribution of stain is not regular around its periphery, probably because of the phenomenon of "capping" to be discussed later, it can be seen that areas that are red fluorescent are also green fluorescent. The surface antigens of this heterokaryon appear to be completely intermixed, when compared to those of early heterokaryons shown previously.

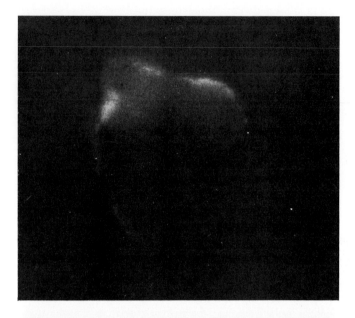

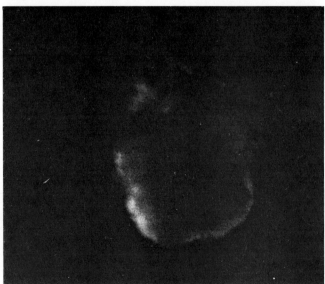

Fig. 1. *Large (4-6 nuclei) Cl ld x VA-2 heterokaryon, stained for mouse H-2 and human HL-A antigens. The antigens are restricted to separate parts of the membrane.*
a. Mouse H-2 antigens (green fluorescence).
b. Human, HL-A antigens (red fluorescence).
250X.

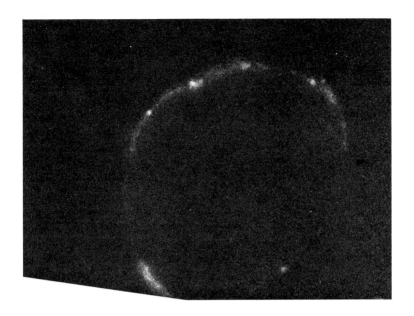

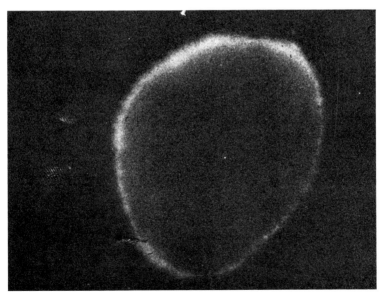

Fig. 2. *3T3 x WI-38 heterokaryon. Mouse H-2 and human species antigens are completely intermixed in this cell.*
a. Mouse H-2 antigens (green fluorescence).
b. Human species antigens (red fluorescence).
600X.

When samples are taken of a population over time, the proportion of segregate and masaic heterokaryons can be estimated for several times after initiating fusion. Data from the original series of experiments in the system are tabulated in Table 1, together with some more recent data on H-2 and HL-A antigens (V. Petit and M. Edidin, unpublished) in later passages of the same cells. While some differences in fraction of segregates surviving at a given time are evident, both columns record a steady and rapid loss of cells with segregated antigens.

This striking loss does not appear to require either synthesis of proteins or glycoproteins or continued generation of ATP. Cells that have been poisoned for up to 6 hours with puromycin or cycloheximide or treated with dinitrophenol are still capable of rapidly intermixing their surfaces when fused. However, intermixing is drastically retarded by lowered temperature. Heterokaryons held for 40 minutes at 15°C (after a few minutes at 37°C to initiate fusion) show a negligible proportion of mosaic cells. While initially this temperature effect was interpreted as due to a phase transition in membrane lipids around 15°, it now appears more likely that the data reflect instead the increasing viscosity of membrane lipids at lowered temperatures. Recently we have been able to observe antigen intermixing in a population of hetero-karyons kept at 15°C; however, the time course of this intermixing was 2-4 times slower than that for cells at 37°C. A similar temperature dependence for diffusion has been reported for rhodopsin (Poo and Cone, 1973).

The observations on: 1) the rapid rate of intermixing of surface antigens, 2) the insensitivity of mixing to inhibitors of protein synthesis and to uncoupling of respiration and 3) the marked effect of temperature on antigen movement, together suggest that this movement is essentially a passive process, rather like diffusion. Indeed, though it would be rash to calculate diffusion constants from the sort of data we can obtain from heterokaryons, a useful estimate can be made from the data of the viscosity of the medium in which diffusion is occurring. This estimate based on:

$$1)\ \eta = \frac{kTt}{6\pi a(\Delta x)^2}$$

is susceptible to variation of 2-4 fold depending upon the values used for a, the radius of the diffusing antigens, t the mean time to diffuse a given distance, and x that distance. a is unknown; though we have fair estimates of the molecular weight of H-2 antigens, nothing is known of their configuration and recent data suggests that they may exist as dimers in the membrane (Schwartz et al., 1973). t is most conveniently taken as the time for half the population to become mosaics, but could be otherwise chosen, and x varies depending upon the cell volume of the parent cells and the heterokaryons. Despite these uncertainties, equation 1 may be used to calculate $1 < \eta < 10$ poise. This is the range of viscosity for many oils, and the calculation

TABLE 1.

Fraction of All Mouse-Human Heterokaryons
Bearing Segregated Antigens

Time after initiating fusion at 37°C	Segregated H-2 and Human Species	H-2 and HL-A
	%	
5	100	–
10	97	72
20	–	45
25	50	–
40	12	28
60	–	25

indicates that by taking antigen intermixing in heterokaryons as due to diffusion we can come up with an estimate of membrane viscosity that approximates our expectations for membrane lipids. Still, it would be desirable to make a better estimate of diffusion by making a series of measurements over time on a single cell. Furthermore, it might also be useful to visualize antigens before beginning the measurements, so that they could be seen directly to spread over the plasma membrane. Such an approach has required the use of rather specialized cells, and modified fluorescent reagents, but it has been achieved, yielding estimates for a diffusion constant for movement of antigens within the plane of the plasma membrane.

Spread of antibody patches on the surface of cultured myotubes

If a small portion only of the surface antigens of a cell could be marked, then, if these antigens were mobile, one would expect to see the mark enlarge with time. Such a system has the advantage of keeping the markers for diffusion in view at all times, and of allowing continuous observation of the marker. However, its use is limited to cells which are large enough themselves to allow marking without reacting the marker with the entire cell surface. Therefore, for these experiments we were forced to leave fibroblasts and to turn to myotubes formed *in vitro* by fusion of fetal rat myoblasts (Yaffe, 1969). In collaboration with Dr. Douglas Fambrough of the Carnegie Institution of Washington, Department of Embryology, myotubes have been marked with spots of fluorescent anti-plasma membrane antibody (prepared against purified muscle plasma membranes), and the rate of enlargement of the spots has been measured from photographs taken at intervals after labelling (Edidin and Fambrough, 1973). Intact antibody itself produces brilliant patches that spread very slowly, perhaps 1 micron per hour. However, if the antibody is first cleaved to monovalent Fab fragments, and then labelled and applied, patch spread approximates a rate of 1 micron per minute. From measurements of this spread, again using a simple approximating formula, we can calculate values for a "diffusion constant" for the marked antigens:

$$2) \; \text{"D"} = \frac{\Delta x^2}{4t}$$

where x = the radius of the patch. "D" is so written because we are dealing with an average for molecules of several molecular weights, and possibly of varying sizes and degrees of insertion in the membrane. We do know that the antigens are protein (they are released from membranes by papain) and that they are of fairly high molecular weight, 10^5 or above based on acrylamide gel electrophoresis of ^{125}I-labelled antigens extracted from intact myotubes. From several fibers, we have derived "D" = $1-2 \times 10^{-9}$ cm^2 sec^{-1} a value that, compared to values for proteins in water (of the order of 10^{-6} cm^2 sec^{-1}) again indicates a viscosity for the myotube plasma membrane of several poise, hundreds of times that of water. The numbers that we obtained range somewhat, but compare well with values independently derived for lipid diffusion in artificial and natural membranes (Devaux and McConnell, 1972; Scandella *et al.*, 1972) and for lateral diffusion of rhodopsin in disc membranes (Poo and Cone, 1974).

Again, they report on a plasma membrane in a fluid state, though the fluid is some hundreds of times more viscous than water.

Perturbation of membrane molecules – Cap Formation in Lymphocytes and L-cell Fibroblasts

Both of the systems that I have described are cumbersome and tedious. However, besides demonstrating motion in the plane of the membrane, they yield numerical estimates for the rate of this motion and this is the chief justification for continuing their use. Still, it would be convenient to develop a method for quickly testing the mobility of membrane proteins, even if the method could not evaluate the degree or rate of this motion. Such a rapid method at first appeared to be offered by the phenomenon of capping (Taylor *et al.*, 1971), or by that of antibody-induced modulation of antigens (Oda and Puck, 1961; Takahashi, 1971; Schlesinger and Chaovat, 1972). Since the latter does not allow viewing of single cells, or localization of their antigens but rather measures loss of cell susceptibility to lysis by antiserum and complement, we have investigated the former, first described by Taylor and colleagues. They reacted lymphocytes with fluorescent antiglobulin antibody. The cells stained in outline when kept cold, but upon warming to 37° all of the stain collected in a cap at one pole of the cell. On the basis of inhibitory effects of dinitrophenol and cytochalasin B, Taylor *et al.* suggested that cap formation depended upon aggregation of mobile membrane antigens by multivalent antibody, followed by collection of aggregates at one pole of the cell. It appeared that antigen aggregation was a passive process, involving

Brownian movement of membrane molecules to a point at which they reacted with antibody and aggregated, while cap formation required cell metabolism, and in lymphocytes seemed to depend upon a cytochalasin B-sensitive cell organelle, perhaps microfilaments. Though capping was temperature sensitive, aggregate formation could continue down to very low temperatures (de Petris and Raff, 1972).

Our investigations used fibroblasts, and generally agreed with those of Taylor *et al.* It appeared that caps would form when fibroblasts, attached to a substrate, were reacted with anti-H-2 antibody and labelled antiglobulin (Edidin and Weiss, 1972). Fibroblasts stained in the cold with these reagents are shown in Figure 3. In Figure 4 are shown cells similarly treated, but incubated at 37°C for 45-60 minutes. It will be seen that stain has left the processes of these cells and accumulated at the cell body. This path of movement strongly suggested that the active event in cap formation, the step requiring cell metabolism, was flow of membrane away from the ruffled edge of a cell. Such a flow has been postulated by Abercrombie and co-workers (1970) as responsible for the rapid transport of particles of metal or carbon centripetally over the cell surface. When fibroblasts were treated with

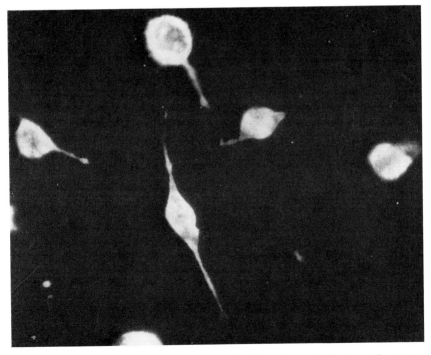

Fig. 3. *Cl ld fibroblasts stained for a single H-2 antigen, H-2.8. Green fluorescence 550X.*

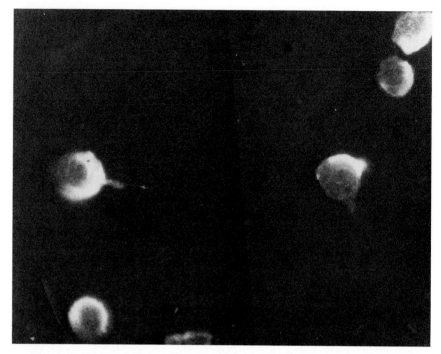

Fig. 4. *As Figure 3, but capping has been induced in these cells by incubation at 37°C for 1 hour. 550X.*

colcemid, destroying their morphological polarisation (Gail and Boone, 1971) cap formation was abolished. While cyctochalsin B had no effect on cap formation in fibroblasts, we were able to confirm the findings of Taylor *et al.* in lymphocytes. Perhaps in those cells morphological polarity is maintained by a filament system, rather than by an arrangement of microtubules, as appears to be the case for fibroblasts.

In any case, we were left with an apparently consistent analysis of cap formation that indicated that if capping occurred in a cell it could be taken as evidence for mobility of the molecules capped, but that failure to cap membrane molecules might as easily be due to absence of membrane flows or ruffling in cells, as to restriction on mobility of membrane proteins. Capping is interesting in its own right, and can be used to alter the cell surface in various ways, but it cannot be relied upon as a rapid general test of mobility in the plasma membrane that we had hoped to devise.

Capping in other fibroblasts and fusion of these cells to form heterokaryons

Though capping could not be generalized as a test for mobility of membrane proteins in all cells, we did expect to induce caps with ease in all

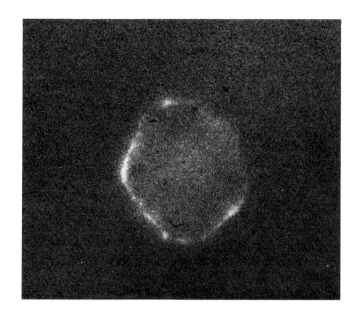

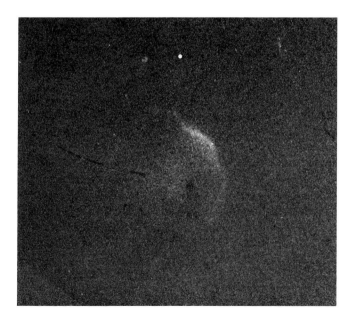

Fig. 5. *3T3 x WI-38 heterokaryon, with segregated H-2 and human species antigens.*
a. Mouse H-2 antigens (green fluorescence).
b. Human species antigens (red fluorescence).
700X.

fibroblasts, especially in those bearing large ruffled membranes. Fibroblast strains prepared in our laboratory from adult mouse kidney, WI-38 human diploid fibroblasts, and the established lines 3T3 and SV40 transformed 3T3, were tested for capping. Of all these, only SV40 3T3 capped. All the others, though they are motile ruffled cells failed to cap either in suspension or on substrates. This finding was totally unexpected, and forced us to consider the possibilities that: 1) the model proposed for the mechanism of capping was wrong or 2) that surface antigens of contact inhibited fibroblasts were restricted in mobility. While the first possibility led to some thought but no improvement of the model, the second suggested the experiment of fusing cells that failed to cap, returning to heterokaryons to measure lateral mobility of surface antigens.

The details and problems of the method of measurement have been discussed previously. A typical result of the method is shown in Figure 5. The cell is a segregate heterokaryon formed between 3T3 and WI-38 parents, stained at two hours after initiating fusion. In Figure 6 are summarized results of a number of fusions of the type shown in Figure 5. The rate of loss of segregates in this cell combination is contrasted with the rate of loss of segregates from a Cl ld x VA-2 population. In other fusions we have combined membrane from a contact inhibited cell with that of a transformed cell (for example, SV40-transformed 3T3 x WI-38) in all cases, presence of one contact inhibited parent in the cross was sufficient to yield heterokaryons that lost segregates slowly compared to our standard Cl ld x VA-2 population.

Rather than stress the apparent contrast between fusion behavior of membranes from contact inhibited compared to transformed cells, I wish to

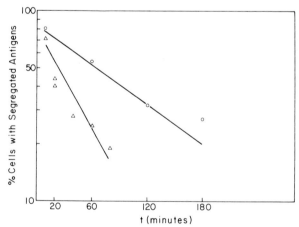

Fig. 6. *Rate of loss of heterokaryons bearing segregated antigens.*
o 3T3 x WI-38
Δ Cl ld x VA-2

emphasize that we have found cells that restrict lateral mobility of membrane proteins, and that can be manipulated *in vitro*. While the demonstration, and especially the quantitative evaluation, of diffusion of membrane protein molecules is worth continuing, it is probably time to begin to examine ways in which this mobility can be restricted. It is clear that in tissues mobility of membrane components is restricted, but it is by no means evident how this is accomplished (Edidin, 1972). The heterokaryon systems that we are now investigating, whatever they may tell us about tumor *versus* normal cells, ought to allow us to work on the more general problem of organization of molecules in the plane of the plasma membrane.

REFERENCES

1. Abercrombie, M. 1961. The bases of locomotory behavior of fibroblasts. *Exp. Cell Res. Suppl. 8*:188-198.

2. Abercrombie, M., J. E. Heaysman and S. Pegrum. 1970. The locomotion of fibroblasts in culture. II. "Ruffling". *Exp. Cell Res. 60*:437-444.

3. Devaux, P. and A. M. McConnell. 1972. Lateral diffusion in spin-labelled phosphatidyl choline multilayer. *J. Am. Chem. Soc. 94*:4475.

4. Edidin, M. 1972. Aspects of plasma membrane fluidity. In "Membrane Research," C. F. Fox, ed., Academic Press, New York, pp. 15-25.

5. Edidin, M. and D. Fambrough. 1973. Fluidity of the surface of cultured muscle fibers. Rapid lateral diffusion of marked surface antigens. *J. Cell Biol. 57*:27-39.

6. Edidin, M. and A. Weiss. 1972. Antigen cap formation in cultured fibroblasts: A reflection of membrane fluidity and of cell motility. *Proc. Nat. Acad. Sci. U.S. 69*:2456-2459.

7. Frye, L. D. and M. Edidin. 1970. The rapid intermixing of cell surface antigens after formation of mouse-human heterokaryons. *J. Cell Sci. 7*:319-335.

8. Gail, M. H. and C. W. Boone. 1971. Effect of colcemid on fibroblast motility. *Exp. Cell Res. 65*:221-227.

9. Gershon, D. and L. Sachs. 1963. Properties of a somatic hybrid between mouse cells of different genotypes. *Nature 198*:912.

10. Gordon, S. and Z. Cohn. 1970. Macrophage-melanocyte heterokaryons. I. Preparation and properties. *J. Exp. Med. 131*:981-1003.

11. Harris, A. 1972. Surface movement in fibroblast locomotion. Acta Protozoologica XI:145-151.

12. Harris, H., E. Sidebottom, D. M. Grace and M. E. Bramwell. 1969. The expression of genetic information: A study with hybrid animal cells. *J. Cell Sci. 4*:499-525.

13. Marcus, P. I. 1962. Dynamics of surface modification in myxovirus-infected cells. *Cold Spr. Harb. Symp. Quant. Biol. 27*:351-365.

14. Oda, M. and T. T. Puck, 1961. The interaction of mammalian cells with antibodies. *J. Exp. Med. 113*:599-610.

15. dePetris, S. and M. Raff. 1972. Distribution of immunoglobulin at the surface of mouse lymphoid cells as determined by immunoferritin electron microscopy. Antibody-induced, temperature dependent redistribution and implications for membrane structure. *Eur. J. Immunol. 2*:523-535.

16. Poo, M-M. and R. A. Cone. 1974. Lateral diffusion of rhodopsin in the photo receptor membrane *Nature 247*:438-440.

17. Porter, K., D. Prescott and J. Frye. 1973. Changes in surface morphology of Chinese hamster ovary cells during the cell cycle. *J. Cell Biol. 57*:815-836.

18. Scandella, P. J., P. Devaux and H. M. McConnell. 1972. Rapid lateral diffusion of phospholipid in rabbit sarcoplasmin reticulum. *Proc. Nat. Acad. Sci. (USA) 69*:2056-2060.

19. Schlesinger, M. and M. Chaovat. 1972. Modulation of H-2 antigenuity on the surface of murine peritoneal cells. *Tissue Antigen 2*:427-435.

20. Schwartz, B. D., K. Kato, S. E. Cullen and S. G. Nathenson. 1973. H-2 histocompatibility alloantigens. Some biochemical properties of the molecules solubilized by NP-40 detergent. *Biochemistry 12*:2157-2164.

21. Takahashi, T. 1971. Possible examples of antigenic modulation affecting H-2 antigens and cell surface immunoglobulins. *Transplant. Proc. 3*:1217-1220.

22. Taylor, R., P. Duffus, M. Raff and S. dePetris. 1971. Redistribution and pinocytosis of lymphocyte surface immunoglobulin molecules infected by anti-immunoglobulin antibody. *Nature (N.B.) 233*:225-229.

23. Watkins, J. F. and D. M. Grace. 1967. Studies on the surface antigens of interspecific mammlian cell heterokaryons. *J. Cell Sci. 2*:193-204.

24. Weiss, L. and K. Clement. 1969. Studies on cell deformability. Some theological considerations. *Exp. Cell Res. 58*:379-387.

25. Wolpert, L. and C. H. O'Neill. 1962. Dynamics of the membrane of *Amoeba proteus* studies with labelled specific antibody. *Nature 196*:1261-1266.

26. Yaffe, D. 1969. Cellular aspects of muscle differentiation *in vitro*. *Current Topics in Dev. Biol. 4*:37-77.

ALTERATION OF PROTEIN COMPONENTS OF TRANSFORMED CELL MEMBRANES IN TWO VIRUS-CELL SYSTEMS

Boyce W. Burge, H. Sakiyama[1] and Gary Wickus[2]

Department of Biology
Massachusetts Institute of Technology
Cambridge, Mass. 02139

INTRODUCTION

Wu *et al.* (1969) and Culp *et al.* (1971) have reported that Swiss mouse 3T3 cells transformed by Simian virus 40 have markedly reduced amounts of both amino and neutral sugars. We examined these same cells and reported that the molecular weights of glycoproteins and glycopeptides of the transformed cells (as determined by acrylamide gel electrophoresis and gel exclusion chromatography) were the same as those of the normal cells (Sakiyama and Burge, 1972). In particular we found that the sialic acid composition of partially purified glycoproteins and glycopeptides was the same for comparable fractions from normal and transformed Swiss 3T3 cells; this result argues against the possibility that differences in the carbohydrate content of normal and transformed cells are due to "incomplete" carbohydrate chains in transformed cells, as was suggested by Hakomori and Murakami for the case of the glycolipids of transformed cells (1968). We were instead forced to conclude that SV40 transformed Swiss 3T3 cells have a lesser number of glycoprotein molecules than do normal cells, possibly due to a reduction in the area of glycoprotein containing membranes of transformed cells.

---

[1] Dr. Sakiyama's current address is:
Division of Physiology and Pathology
National Institute of Radiation Science
4 Anagawa, Chiba-shi
Chiba, Japan

[2] Dr. Wickus is a postdoctoral fellow of the Damon Runyon Fund for Cancer Research. The work reported here concerning Rous sarcoma virus transformation of chicken cells was carried out by Dr. Wickus in the laboratory of Dr. P. W. Robbins.

To establish the generality of this phenomenon we have extended our studies to 3T3 cells from another strain of mice, the BALB/C strain, and to a second isolate of Swiss 3T3 cells. In the BALB/C system the depression of sialic acid in transformed cells is even more marked than in the Swiss 3T3 system studied by Wu *et al.* (1969), SV40 transformed BALB/C 3T3 cells have only 35% the sialic acid of their normal counterparts (Grimes, 1970). The results reported here confirm our original conclusions: in all cases the sialic acid content of the transformed cell was markedly reduced without detectable change in the molecular weight or sialic acid content of isolated glycopeptides.

In another cell-virus system one of us (G. Wickus) has examined changes in cell protein on transformation, using a temperature sensitive mutant of Rous sarcoma virus, Ts-68, isolated by Kawai and Hanafusa (1971). This mutant has the useful property of transforming primary or secondary chick embryo fibroblasts rapidly at 36°C so that the transformed cells are indistinguishable from cells transformed by wild type Schmidt-Rupin Rous sarcoma virus. However, when Ts-68 infected cells are shifted to 41°C they regain a normal untransformed phenotype within 16 hours. Infectious virus production is the same at 36° and 41°, and the temperature sensitive transformation process is fully reversible. The existence of such a mutant is strong evidence for a virus function which is necessary both for the initial transformation event and for the continued maintenance of transformation. Using this sytem, and making use of pulse labeling with ^{35}S-L-methionine, and surface labeling of cells with ^{125}I, it was possible to demonstrate several distinct changes in the acrylamide gel electrophoresis profiles of proteins from cells in the process of transformation.

ABBREVIATIONS

EDTA: ethylendiaminetetraacetic acid
B.D.: blue dextran
P.R.: phenol red
TCA: trichloroacetic acid
PBS: phosphate buffered saline
SDS: sodium dodecyl sulfate
CEF: chick embryo fibroblast
SR-RSV-A: Schmidt-Rupin strain of Rous sarcoma virus, subgroup A
Ts-68: thermosensitive mutant of SR-RSV-A

EXPERIMENTAL PROCEDURES

Cells and medium:

3T3 Cells: BALB/C 3T3 cells (clone A31), SV40 transformed A31 cells (SVT2 cells), Swiss 3T3 cells (clone 3T3-M) and SV40 transformed 3T3-M cells (SV3T3 cells) were obtained from Dr. George Todaro and Dr. Howard

Green, respectively. The saturation density of A31 and 3T3-M was 1×10^6 and that of SVT2 and SV3T3 was 1×10^7 cells per 50 mm petri plate, respectively. The medium used for all experiments was Eagle's minimal essential medium with four times the normal concentration of vitamins and amino acids, 10% fetal calf serum, and penicillin 75 units/ml and streptomycin 50µg/ml.

Chick embryo fibroblasts and Ts-68 infection: Chick embryo fibroblasts (CEF) were prepared from 11-day embryos (COFAL-negative, c/o, SPAFAS, Norwich, Connecticut) as described by Rein and Rubin (1968). Primary cells were seeded at a density of 1×10^7 cells per 100 mm plastic Petri dish (Falcon) in medium 199 supplemented with 1% calf serum, 1% heat inactivated chick serum and 2% tryptose phosphate broth. After 3 hours at 39°C, medium was removed from cultures to be infected and 1 ml of virus added per culture (0.5-1.0×10^7 focus forming units per ml). After 1 hour, medium described above was added. For preparation of secondary cultures, cells were trypsinized and transferred at 5×10^6 cells per 100 mm Petri plate in medium 199 supplemented with 4% calf serum, 1% heat-inactivated chick serum and 10% tryptose phosphate broth. Other procedures were as described by Wickus and Robbins (1973).

Labeling and harvest of cells:

Cells to be labeled with isotopes of amino sugars were grown in roller bottles and on reaching confluence were labeled with either ^{14}C-glucosamine or ^{3}H-glucosamine for 2 days. All experiments were carried out in duplicate, with reversal of the ^{3}H and ^{14}C labels. Harvesting of labeled monolayers was begun by washing three times with phosphate buffered saline (PBS). Cells were then incubated with TNE buffer (0.5M Tris, 0.1M NaCl, 0.01M EDTA pH 7.4) for five minutes at 37°C. Cells were then broken in a glass Dounce homogenizer in half the normal concentration of cold reticulocyte standard buffer (0.005 NaCl, 0.005M Tris pH 7.5, 0.001M $MgCl_2$). The fractionation of the membrane, the preparation of samples for gel electrophoresis and the pronase digestion were carried out as described by Sakiyama and Burge (1972).

Label procedures for Ts-68 infected CEF are described in figure legends.

Chromatography of glycopeptides:

Chromatography of pronase digested glycopeptides by Biogel P-10 was carried out on a 1×105 cm column which was eluted with 0.1M sodium phosphate buffer, pH 7.8 containing 0.1% SDS, 0.1% mercaptoethanol and 0.01% EDTA. Fractions of 0.8 ml were collected at a rate of 9.5 ml per hour. Glycopeptides of subcellular fractions were analyzed by DEAE-cellulose column chromatography. The column was eluted with a linear gradient of NaCl (0.0-0.6M in 0.0025M sodium phosphate buffer pH 7.6). The flow rate was maintained at 23 ml per hour. Fraction volume was 6 ml. Uronic acid was determined by the method of Bittner and Ewin (1961).

Gel electrophoresis:

These procedures were carried out as described by Sakiyama and Burge (1972). The Laemmli method (Laemmli 1970) of acrylamide gel electrophoresis was used for studies of chick embryo fibroblast proteins, and adapted to slab gels and autoradiography.

RESULTS

Analysis of glycoproteins by acrylamide gel electrophoresis.

EDTA release surface fraction:

In Figure 1 is shown the effect of EDTA on the release of glucosamine labeled components from the cell surface as a function of incubation time. The release of trichloroacetic acid (TCA) insoluble radioactivity occurred rapidly for 5 minutes after the addition of EDTA, and then leveled off. This behavior is consistent with a surface location for released material. Additional evidence for a surface location of this material has been presented previously (Sakiyama and Burge, 1972). The EDTA extract was centrifuged at 500g for 10 minutes to remove cells and the supernatant fraction was layered on a discontinuous sucrose gradient which was centrifuged at 25,000 rpm for 3 hours, at 5°C. (see Sakiyama and Burge, 1972).

The discontinuous gradient was made up of three sucrose solutions, 1.10, 1.18 and 1.20 g/m^3 respectively. The distribution of TCA insoluble radioactivity on the gradient was 30% in the sample volume, 65% in the solution of density 1.10 g/m^3 and 3% at the interface of the 1.10 g/m^3 and 1.18 g/m^3 solutions. The result suggests that most of the removed glycoprotein molecules are not present as true solutes (in which case they would remain in the sample volume) but are in small aggregates, perhaps vessicles. Culp and Black (1973) have in fact suggested that EDTA chelation may cause budding of vessicles from the surface membrane.

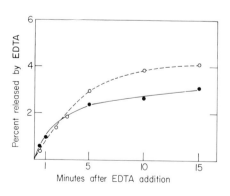

Minutes after EDTA addition

Fig. 1. *The release of glucosamine labeled components from the cell surface by EDTA: Confluent cells were labeled with ^3H glucosamine for 48 hours. Cells were then washed with warm PBS and incubated with 0.01M EDTA buffered with 0.5M TRIS pH 7.4 and containing 0.1M NaCl at 37°C. Cells that came off the plate during this treatment were centrifuged and combined with the cells which did not come off the plate. Cells were dissolved with 1 ml of 1N NaOH. TCA insoluble radioactivity was determined both in EDTA extracts and cells. ●——● SVT2 cells ○——○ A31 cells.*

A comparison of EDTA released glycoproteins of A31 and SVT2 cells is shown in Figure 2A. The most striking difference was found in peaks E-2 and E-3. The profile of SVT2 cells shows a prominent peak E-2 and smaller peak E-3. In the case of A31 cells, peak E-2 is missing and peak E-3 is more marked than in the fractions from SVT2 cells. These differences were also observed when the ^{3}H and ^{14}C labels were reversed. However, when the Swiss cells (3T3-M and SV3T3) were examined, no difference in the relative amounts of peak E-2 material was observed. Therefore, the difference in amount of peak E-2 material seen in the comparison of the BALB/C lines, A31 and SVT2, may be a result of selection after transformation and may not reflect any general difference between normal and transformed 3T3 lines. The EDTA extracts were directly compared to membrane fractions by co-electrophoresis of the appropriate fractions. Peak E-3 co-migrates with peak M-2 (membrane) and peak E-2 co-migrates with the small shoulder M-2 at fraction 12, Figure 2B.

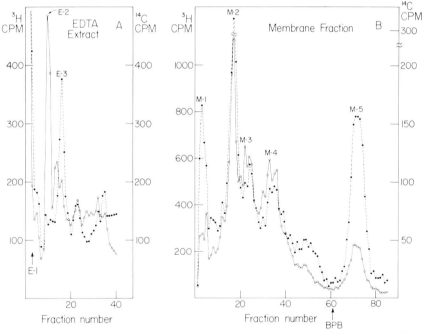

Fig. 2. *Comparison of glycoproteins from normal and transformed cells: EDTA extracts and membrane fractions on polyacrylamide gels. A31 and SVT2 cells were labeled at confluence with* ^{14}*C-glucosamine and* ^{3}*H-glucosamine respectively, for 48 hours. Glycoproteins of EDTA extracts and membrane fractions were solubilized with 2% sodium dodecyl sulfate (SDS) 0.5 M urea, 100 mM dithiothreitol (DTT) and heated at 90°C for 5 min. Gels were 7.5% acrylamide, 9 cm in length and 0.6 cm in diameter. Electrophoresis was for 15 hours at 3 mA per gel at room temperature. Migration is from left to right.* ○———○ ^{3}*H-glucosamine* ●———● ^{14}*C glucosamine.*

Membrane fraction:

Figure 2B is a comparison of the glycoproteins of the membrane fractions of A31 and SVT2 cells. These profiles consist of five major peaks. Peak M-1, partially excluded from the gel is chiefly mucopolysaccharide while M-5 is for the most part glycolipid. The remaining peaks which reflect glycoprotein migration, reveal no difference in the position of migration of glycoprotein from the two cell lines. However, the amount of peak M-2 relative to peaks M-3 and M-4 was found to be higher in A31 cells than in SVT2 cells. When Swiss lines (3T3-M and SV3T3) were examined by the same techniques identical results were found. This reduction in the amount of peak M-2 in transformed cells relative to normals is most pronounced when the normal cells used for comparison have a very low saturation density.

It is of interest that for all of the six cell lines we have examined in this and in an earlier report (Sakiyama and Burge, 1972) peak M-2 has identical mobility on acrylamide gels.

Analysis of glycopeptides by Biogel P-10 exclusion chromatography.

Glycopeptides were produced from glycoprotein fractions (either crude or electrophoretically purified) by pronase digestion. The size distribution and sialic acid content of these glycopeptides was then determined by exclusion chromatography on Biogel P-10. Whole EDTA extracts and whole membrane fractions of A31 and SVT2 cells were compared in this way (Fig. 3A, 3B).

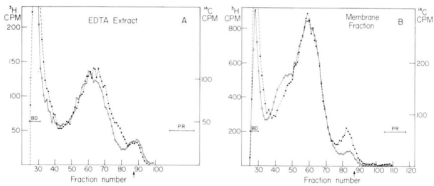

Fig. 3A. *Comparison of glycopeptides from normal and transformed cells: EDTA extracts and membrane fractions on Biogel P-10. SVT2 cells were labeled with ^{14}C-glucosamine and ^{3}H-glucosamine respectively. The label was reversed in Figure 3A. EDTA extracts and membrane fractions were digested with pronase and the resulting glycopeptides were fractionated on Biogel P-10. The column was eluted with 0.1 M sodium phosphate buffer, pH 7.8 containing 0.1% SDS, 0.1% mercaptoethanol and 0.01% EDTA. Samples of 0.8 ml were collected at a rate of 9.5 ml per hour. BD represents the samples in which blue dextran was eluted: PR represents the samples into which phenol red marker was eluted. Figure 3A ○———○ ^{14}C-glucosamine ●———● Figure 3B ^{3}H-glucosamine ○———○ ^{3}H-glucosamine ●———●———● ^{14}C-glucosamine.*

The small peak eluting prior to phenol red (arrow) is presumably glucosamine or galactosamine since this peak migrates between sialic acid and sucrose. We found no significant differences in the glycopeptide profiles of A31 and SVT2 cells on Biogel P-10 with either EDTA extracts or membrane fractions, with the exception of a marked shoulder seen on the high molecular weight side of the main glycopeptide peak of the membrane fraction from SVT2 cells. This shoulder may correspond to that seen by Buck et al. (1970) in several transformated cell lines. This shoulder was also seen in our studies of Swiss 3T3 cells (1972).

Glycoproteins from peak M-2 of normal and transformed cells were eluted from the gel, digested with pronase and compared on Biogel P-10 (Figure 4). The small peak eluting just behind blue dextran was found in greater amounts in SVT2 than A31 cells. The area of this peak decreased after the glycopeptides were hydrolyzed with 0.1N H_2SO_4, and radioactivity appeared at the position characteristic of sialic acid. The percentage of radioactivity found in the sialic acid peak was 15% and 12% for SVT2 and A31 cells, respectively. The sensitivity of Biogel P-10 chromatography to molecular weight change can be roughly gauged by noting that the major glycopeptide peak in Figure 4 shifts four fractions (toward a lower molecular weight) after removal of sialic acid with H_2SO_4. Glycopeptides from peak M-3 were also analyzed on Biogel P-10 chromatography. The profiles of the glycopeptides of peak M-3 and A31 and SVT2 were identical. We then examined also the glycopeptides derived from the glycoprotein peaks E-2 and E-3 of the EDTA extracts. Glycopeptides from these peaks were more homogeneous by size than those from the whole EDTA fraction, indicating a narrower spectrum of molecular weights (result not shown). These profiles were in fact very similar to the glycopeptide

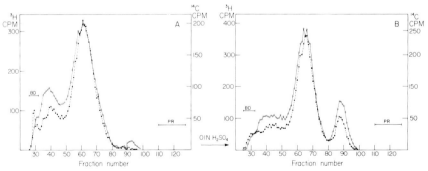

Fig. 4. *Comparison on Biogel P-10 of glycopeptides from the glycoproteins of peak M-2: Glycoproteins from peak M-2 of the membrane fraction were eluted from acrylamide gel and digested with pronase. The resulting glycopeptides were fractionated on Biogel P-10.*
A: *Before hydrolysis*   B: *After hydrolysis*
o———o *^{3}H-glucosamine (SVT2)*   ●———● *^{14}C-glucosamine (A31)*

profiles of the M-2 membrane glycoprotein of these cells, shown in Figure 4A. Again, there were no significant differences in the profiles of the glycopeptides from these peaks of A31 and SVT2 by this method of analysis (Results are not shown). From these combined results we infer that no major shift in the size distribution of glycoproteins and glycopeptides has occurred.

Analysis of glycopeptides by DEAE cellulose chromatography.

Grimes (1970) has reported that SVT2 cells have only 35% as much sialic acid as A31 cells (i.e. micrograms sialic acid per milligram protein). If there were present in transformed cells glycopeptides of the same structure as those found in normal cells, differing only in that they lack terminal sialic acid, these glycopeptides should be separable by DEAE cellulose column chromatography, where separation is based principally on electrical charge. Therefore glycopeptides from EDTA extracts and membrane fractions were analyzed by DEAE cellulose column chromatography; glycopeptides from the EDTA fraction are shown in Figure 5A, glycopeptides from the membrane fraction are shown in Figure 5B. The recovery of radioactivity was about 80% in every case. Both the EDTA extracts and the membrane fractions contained fairly large amounts of radioactivity which did not attach to DEAE cellulose. In the case of membrane fractions from A31 and SVT2 cells 42.6% and 42.0% of the radioactivity failed to bind to the column. The profiles of material from A31 and SVT2 cells were identical except that in both EDTA extracts and membrane fractions peak 4 was found in greater amounts in A31 than in SVT2 cells. The profiles of glycopeptides of EDTA extracts appeared less complex than those of membrane fractions again suggesting that the EDTA extract is a specific subfraction of the cell surface.

To examine how mucopolysaccharides behave on DEAE cellulose, umbilical cord hyaluronic acid (Sigma) was mixed with glycopeptides from both EDTA extracted material and membrane fraction and chromatographed on DEAE cellulose (Figure 5C). The hyaluronic acid peak was found to elute from the DEAE cellulose with peak 3. Thus we conclude that peak 3 is probably hyaluronic acid. It should be noted that there is much more hyaluronic acid in the EDTA surface fraction than in the membrane fraction (Figure 5A, B).

Analysis of the proteins of Ts-68 transformed CEF

Chick embryo fibroblasts at 41° or 36°C display a characteristic parallel arrangement although they do not have the exquisite contact inhibition of growth shown by 3T3 mouse fibroblasts. In Figure 6 are seen photomicrographs of secondary CEF infected with Ts-68. At 41° (A) the cells have the parallel arrangement and morphology of uninfected CEF. Within 1.5 hours after the shift to 36° cells become more refractile (B), and increasingly so at 3

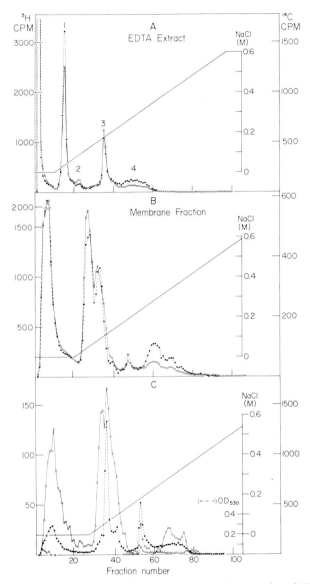

Fig. 5. *DEAE cellulose column chromatography of glycopeptides of EDTA extracts and membrane fractions. A31 and SVT2 cells were labeled with ^{14}C-glucosamine and ^3H SVT2 cells respectively. Glycopeptides were analyzed on DEAE cellulose column. Column size was 1.7X15 cm. The sample was applied in 0.0025 M sodium phosphate buffer pH 7.6 and a linear gradient of NaCl from 0 to 0.6 M in 0.0025 M phosphate buffer pH 7.6 was employed. Hyaluronic acid used as a marker was determined by colorimetric tests* o——o ^3H-glucosamine ●——● ^{14}C-glucosamine △——△ DD_{530}

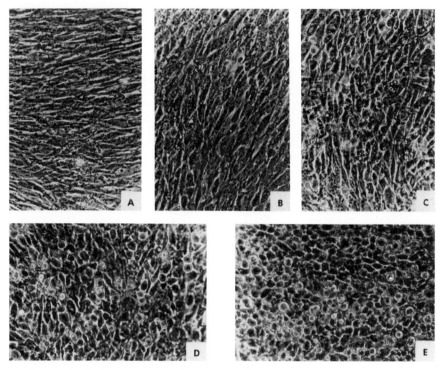

Fig. 6. *Shift of Ts-68 infected CEF monolayers from 41° to 36°C. The cells were grown as described in Experimental Methods*
  A) *Cells at 41°*
  B) *1.5 hours after shift to 36°*
  C) *3 hours after shift to 36°*
  D) *11 hours after shift to 36°*
  E) *24 hours after shift to 36°*

hours after shift (C). A fully developed transformed morphology is seen by 11 hours (D) and 24 hours (E) after temperature shift. The transformed monolayer is characterized by round refractile cells. The sequential change in morphology is fully reversible on shift back to 41°, and infectious virus is produced in equal amounts at either 41° or 36°C. Production of virus at both temperatures is an important consideration, since it might be suspected that differences in protein composition of Ts-68 infected cells at 41° and 36° might reflect different amounts of virus-specified protein at the two temperatures. In fact, it is only with great difficulty that one detects SR-RSV-A structural proteins in infected cells. In an earlier publication (1973) Wickus and Robbins reported that a protein with the same electrophoretic mobility as the major group specific antigen of SR-RSV-A is present in small amount in plasma membrane preparations of Ts-68 infected cells at both 36° and 41°. Thus the major group specific antigen (~30,000 daltons) appears to be synthesized in equal amounts at both 36° and 41°C.

In Figure 7 is recorded the result of an experiment in which cells were shifted from the non permissive (41°) to the permissive temperature (36°) and pulsed with high specific activity ^{35}S-L-methionine for 3 hour periods both before (A,B) and after (C→J) the shift to 36°. At each labeling interval duplicate petri plates were rinsed with PBS, disrupted with SDS, and examined by Laemmli's method (1970) of polyacrylamide gel electrophoresis and autoradiography. The labeling intervals presented are 3 hours before the temperature shift (A,B) and 0-3h. (C,D) 3-6h. (E,F) 6-9h. (G,H) and 9-12h. (I,J) after the shift from 41° to 36°. The two most striking changes are indicated by the Roman numerals I and II in Figure 7. As indicated by the arrow at I, a high molecular weight protein of ~100,000 daltons (the lower band of a doublet) present before the shift is seen to vanish from the gels

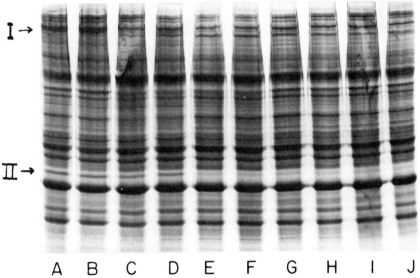

$I \rightarrow$

$II \rightarrow$

A   B   C   D   E   F   G   H   I   J

Fig. 7. *Autoradiogram of an 8% polyacrylamide slab gel electropherogram of CEF infected with Ts-68. Monolayers were labeled with ^{35}S-L-methionine at 3 hour intervals following a shift from the nonpermissive (41°) to permissive temperature (36°). The complete monolayers were removed by disrupting with SDS and scraping with a rubber policeman. Phenylmethane sulfonyl fluoride was added prior to disruption to minimize possible proteolysis. Electrophoresis is from top to bottom.*

A,B) 3 hour pulse at the nonpermissive temperature (41°)
C,D) 3 hour pulse immediately after shift from 41° to 36°
E,F) 3 hour pulse from 3-6 hours after shift to 36°
G,H) 3 hour pulse from 6-9 hours after shift to 36°
I,J) 3 hour pulse from 9-12 hours after shift to 36°

during the first 3 hours after shift to 36° (C,D). This protein (component I) is either no longer synthesized after shift down, or it is degraded rapidly or otherwise removed from cells, and thus does not appear on the gels. In the same 3 hour interval immediately following shift to 36°, a new protein appears on the gels migrating about 1 mm ahead of the vanished component I (C,D) and this new protein increases somewhat during the next interval (3-6 hours after shift down, (E,F)). Current experiments are directed at determining whether this newly appearing protein is a proteolytic product of component I, or a unique induced species. Of great interest is the observation that the changes described above occur almost immediately after cells are shifted to the transformation—permissive temperature; these are perhaps the earliest described events occurring in transformation and may help to shed light on the molecular mechanism of the transformation.

The component II indicated by arrow in Figure 7 is a protein of approximately 45,000 daltons and migrates only slightly more slowly than a prominent ^{35}S-L-methionine-labeled band of 43,000 daltons which itself comigrates with chick muscle F-actin. Component II is labeled during the first 3 hour interval after shift to 36° (C,D) but during the next 3 hours (E,F) and subsequent intervals (G,H,I,J) it is not synthesized, or fails to accumulate in the cell for some other reason. Though equally striking, this change occurs subsequent to the change observed in component I, since it is not immediately manifest after shift to 36° (e.g. little change in gels C,D). It is thus a secondary event in transformation.

The 43,000 dalton species referred to above, which has been found to co-migrate with F-actin, has been found in reduced amounts in membrane preparations of SR-RSV-A transformed cells (Wickus and Robbins, 1973). This protein is also found in reduced amounts in membrane preparations from Ts-68 infected cells incubated at 36° (the permissive temperature for transformation); the effect is seen both with vessicles prepared by the method of Perdue et al. (1971) or cell ghosts prepared by the method of Brunette and Till (1971). This is a significant observation since McNutt et al. (1973) have demonstrated a striking morphological difference between the membrane associated α filaments of 3T3 and SV40 transformed 3T3 mouse fibroblasts; the α filaments, which are shown to have actin-like activity in reactions with heavy meromyosin, are present as a dense, highly organized network immediately beneath the surface of the plasma membranes of 3T3 cells. In contrast, the SV40 transformed 3T3 cells have a diffuse and poorly organized system of membrane-associated α filaments (McNutt et al., 1973). Because of the probable involvement of these actin-like filaments in membrane movements, this phenomena will receive careful experimental attention.

Although controls are not shown here, all the changes observed on shift of Ts-68 infected cells from 41° to 36° are seen in cells infected by wild type

SR-RSV-A at all temperatures between 36° and 41°. Also, by gel electrophoresis criteria Ts-68 infected CEF at 41° are identical to uninfected cells save for the small amount of group specific antigen seen associated with membranes of Ts-68 CEF. Thus the changes described are not artifacts of either the temperature shift or SR-RSV growth *per se*.

Cell surface iodination

Lactoperoxidase catalyzes transfer of ^{125}I to accessible tyrosines to produce monoiodotyrosine. A variety of control experiments with intact red blood cells and enveloped virus have shown that only proteins on the outer face of a membrane are accessible to the lactoperoxidase catalized iodination, when the reaction is carried out under controlled conditions (see Sefton *et al.*, 1973 for the method currently in use in our laboratory). It is of interest that the principal iodinatible CEF surface protein is a species with electrophoretic mobility similar to that of component I in Figure 7 (Wickus *et al.*, 1973). This iodinatible species is readily removed from cells by minute amounts of trypsin (5 μg/ml for 15 min. at 39°) and once removed requires some hours before it reappears as a species susceptible to iodination. SR-RSV-A transformed cells are similar to trypsinized cells in having very little of this 'component I-like' iodinatible band. Further experiments will be required to determine whether this iodinatible material is identical to the component I identified in Figure 7, or whether it is a unique material with electrophoretic mobility coincidently similar to component I. Efforts are being made to improve the resolution of gels in the region of high molecular weight proteins and glycoproteins to help resolve the above uncertainties.

DISCUSSION

SV40 transformed 3T3 fibroblasts

We conclude from this work that there are no differences in the molecular weights of either the glycoproteins or the glycopeptides of normal and transformed cells sufficient to explain the depression of carbohydrate content seen in transformed cells (Wu *et al.*, 1969; Culp *et al.*, 1971; Grimes, 1970). DEAE cellulose column chromatography, a separatory method based on electrical charge, also reveals no difference in the quantities of normal and transformed cell glycopeptides. This is quite striking, since the principal charged residue of glycopeptides, sialic acids is depressed by 65% in transformed BALB/C 3T3 cells. Therefore we suggest that the reduced carbohydrate composition of transformed cells must be due to an absolute decrease in the number of glycoprotein molecules per cell or per milligram protein. This decrease might be accounted for either in terms of a lower density of

glycoprotein per unit area of transformed cell membrane, or by a corresponding decrease of the total membrane area of the transformed cells.
At present we are gathering data to decide this issue.

There are two additional points which should be made:

1) Our results do not rule out changes in the internal structure of the glycopeptides of transformed cells, as, for instance, substitution of one sugar for another, or a change in linkage. They do seem to rule out changes in the molecular weights of a majority of the glycopeptide moieties.

2) In studies of the surface properties of normal and transformed cells one must be wary of fluctuations caused by the aneuploidy of culture cells and consequent genetic drift. For instance, the alteration of the relative amounts of the EDTA released peaks E-2 and E-3, seen in BALB/C 3T3 transformed cells, is not seen in Swiss 3T3 transformed cells. Further, much variability in glycolipid composition is found in clones derived from the Nil 2 cell line (Sakiyama et al., 1972) and this variability is independent of both morphology and tumorigenicity (Sakiyama and Robbins, 1973). At present we are attempting to relate changes seen in the 3T3, SV-40-3T3 system to the alterations observed in the Ts-68-CEF system, with particular attention to changes in component I and the high molecular weight iodinatible species.

Ts-68 infected CEF

The observation of changes in membrane proteins of Ts-68 infected CEF under transformation-permissive and non-permissive conditions is obviously incomplete. Observed changes in components I and II might be due to selective proteolysis (such as that described by Unkeless et al, 1973 and Ossowski et al. 1973), preferential loss of these components into the medium, or an abrupt cessation of synthesis. Preliminary experiments tend to favor the latter suggestion but these experiments are far from complete. At present the experimental strategy is to concentrate on those changes that occur immediately after shift of Ts-68 infected cells to 36°C, and try to determine the identity and possible functions of the proteins involved.

With the current limited data, a synoptic hypothesis would be premature. For those whose primary concern is developmental biology the most important principle to be gleaned from this work is that changes in cell morphology and behavior are likely to be accompanied by alterations in the protein components of the altered cell, particularly the components of the cell membranes. When careful controls are possible (as in a temperature-sensitive system) and methods for separation with high resolution are available, as with the Laemmli gel system, then alterations may be detected at the molecular level. Once alterations are well characterized at the level of protein species with known function, it should be possible to explain differentiation in terms of these changes.

SUMMARY

We have examined two virus-transformed cell systems for membrane protein alterations correlated with transformation and malignancy. In the first of these systems (3T3 mouse fibroblasts transformed by SV40 virus) previous work had demonstrated a pronounced reduction in amounts of both neutral and amino sugars in transformed 3T3 cells. We found however, that the molecular weight distribution of membrane glycoproteins and glycopeptides of both normal and transformed cells was substantially the same. In addition, the relative sialic acid content of glycoprotiens and glycopeptides from the normal and transformed lines was the same. Hakomori and Murakami have suggested (1968), on the basis of work with glycolipids of transformed hamster cells that transformed cells might be characterized by 'incomplete' or shortened carbohydrate chains in glycosylated macromolecules. Our work strongly suggests that this phenomenon does not occur in the carbohydrate chains of transformed cell glycoproteins, though more subtle changes are not ruled out.

A principal objection to comparisons of normal and transformed cells in the case of cell lines is the known aneuploidy of continuous lines and the tendency toward genetic drift under the twin pressures of unstable chromosome complement and continuous passaging *in vitro*. These objections can be overcome by using a system in which a diploid population of cells can be rapidly converted from the normal to the transformed phenotype by use of virus mutants with a temperature sensitive transforming function. One of us (G. Wickus) has used the Ts-68 mutant of the Schmidt Rupin strain of Rous sarcoma virus to transform secondary chick embryo fibroblasts. Fibroblasts so infected have normal morphology at 41°C but are rapidly converted to a transformed morphology on shift down to 36°C. Several striking changes occur soon after shift down: 1) the net synthesis of a protein of greater than 100,000 daltons is immediately reduced. At the same time the net synthesis of a protein of slightly lower molecular weight is sharply increased. 2) the net synthesis of a protein of 45,000 daltons, found in plasma membrane preparations, is reduced, but only after 3 hours at 41°. Other interesting changes also occur on shift down, and all changes in protein patterns are fully reversible when cells are shifted back to 41°C. In addition, all alterations seen in Ts-68 transformed cells at 36° are seen in cells transformed by wild type SR-RSV-A and incubated at either 36° or 41°.

In studies of both SV40 transformed 3T3 cells and RSV-Ts-68 transformed chick embryo fibroblasts it was evident that the great majority of cell proteins are unaffected by transformation. Only by means of high resolution acrylamide gel electrophoresis with which it was possible to examine discrete and well separated populations of proteins did we find evidence of transformation-related alterations.

ACKNOWLEDGEMENT

We wish to acknowledge the skillful technical assistance of D. Holleman, S. Weinzierl, and H. Samuelsdottir. We thank Dr. H. Hanafusa for providing SR-RSV and its temperature sensitive derivative.

This work was supported by grants from the National Institutes of Health: AID 5 RPL AI0 9715-02 and NCI CA 12174-02 to B. W. Burge and 5-R01-AM-06803-10 to P. W. Robbins. Dr. Wickus is a fellow of the Damon Runyon Memorial Fund for Cancer Research.

REFERENCES

1. Bittner, T. and Ewin, R. 1961. A Modified Carbohydrate Reaction for Uronic Acids. *Biochem. J. 81*, 43p.

2. Brunette, D. M. and Till, J. E. 1971. A Rapid Method for the Isolation of L-Cell Surface Membranes Using an Aqueous Two-Phase Polymer System. *J. Membrane Biol. 5,* 215-224.

3. Buck, C. A., Glick, M. C. and Warren, L. 1970. A Comparative Study of Glycoproteins from the Surface of Control and Rous Sarcoma Virus Transformed Hamster Cells. *Biochemistry 9*, 4567-4576.

4. Culp, L. A. and Black, P. H. 1972. Release of Macromolecules from BALB/C Mouse Cell Lines Treated with Chelating Agents. *Biochemistry 11*, 2161-2172.

5. Culp, L. A., Grimes, W. J. and Black, P. H. 1971. Contact-Inhibited Revertant Cell Lines Isolated from SV40 Transformed Cells I. Biologic, Virologic and Chemical Properties. *J. of Cell Biol. 50*, 682-690.

6. Grimes, W. J. 1970. Sialic Acid Transferases and Sialic Acid Levels in Normal and Transformed Cells. *Biochemistry 9*, 5083-5092.

7. Hakomori, S. and Murakami, W. T. 1968. Glycolipids of Hamster Fibroblasts and Derived Malignant-Transformed Cell Lines. *Proc. Nat. Acad. Sci. U.S.A. 59*, 254-261.

8. Kawai, S., and Hanafusa, H., 1971. The Effects of Reciprocal Changes in Temperature on the Transformed State of Cells Infected with a Rous Sarcoma Virus Mutant. *Virology 46*, 470-479.

9. Laemmli, U. K. 1970. Cleavage of Structural Proteins During the Assembly of the Head of Bacteriophage T4. *Nature 227*, 680-685.

10. McNutt, N. S., Culp, L. A. and Black, P. H. 1973. Contact-Inhibited Revertant Cell Lines Isolated from SV40 Transformed Cells. IV. Microfilament Distribution and Cell Shape in Untransformed, Transformed and Revertant BALB/C 3T3 Cells. *J. Cell Biol. 56*, 412-428.

11. Ossowski, L., Unkeless, J. C., Tobia, A., Quigley, J. P., Rifkin, D. B., and Reich, E. 1973. An Enzymatic Function Associated with Transformation of Fibroblasts by Oncogenic Viruses. Part II. *J. Exp. Med. 137*, 112-126.

12. Perdue, J. F., Kletzien, R., and Miller, K., 1971. The Isolation and Characterization of Plasma Membrane from Cultured Cells. I. The Chemical Composition of Membrane Isolated from Uninfected and Oncogenic RNA Virus-Converted Chick Embryo Fibroblasts. *Biochem. Biophys. Acta, 249*, 419-434.

13. Phillips, D. R. and Morrison, M. 1973. Changes in Accessibility of Plasma Membrane Protein as the Result of Tryptic Hydrolysis. *Nature New Biology 242*, 213–215.

14. Rein, A. and Rubin, H. 1968. Effect of Local Cell Concentrations Upon the Growth of Chick Embryo Cells in Tissue Culture. *Exp. Cell Res. 49*, 666-678.

15. Sakiyama, H. and Burge, B. W. 1972. Comparative Studies of the Carbohydrate-Containing Components of 3T3 and SV40 Transformed 3T3 Mouse Fibroblasts. *Biochemistry 11*, 1366-1377.

16. Sakiyama, H., Gross, S. K. and Robbins, P. W. 1972. Glycolipid Synthesis in Normal and Transformed Hamster Cell Lines. *Proc. Nat. Acad. Sci. U.S.A. 69*, 872-876.

17. Sakiyama, H. and Robbins, P. W. 1973. Glycolipid Synthesis and Tumorigenicity of Clones Isolated from the Nil 2 Line of Hamster Embryo Fibroblasts. *Fed. Proc. 32*, 86-90.

18. Sefton, B. M., Wickus, G. G. and Burge, B. W. 1973. Enzymatic Iodination of Sindbis Virus Proteins. *J. of Virology 11*, 730-735.

19. Unkeless, J. C., Tobia, A., Ossowski, L., Quigley, J. P., Rifkin, D. B., and Reich, E. 1973. An Enzymatic Function Associated with Transformation of Fibroblasts by Oncogenic Viruses. Part I. *J. Exp. Med. 137*, 85-111.

20. Wickus, G. G., and Robbins, P. W. 1973. Plasma Membrane Proteins of Normal and Rous Sarcoma Virus-Transformed Chick-Embryo Fibroblasts. *Nature New Biology 245*, 65-67.

21. Wickus, G. G., Branton, P. E. and Robbins, P. W. 1973. Rous Sarcoma Virus Transformation of the Chick Cell Surface. Cold Spring Harbor Symp. Quant. *Biol. 38* (in press).

22. Wu, H. C., Meezan, E., Black, P. H., and Robbins, P. W., 1969. Comparative Studies on the Carbohydrate-Containing Membrane Components of Normal and Virus-Tranformed Mouse Fibroblasts. *Biochemistry 8*, 2509-2517.

SESSION IV
Transformation of Cells in Culture

Moderator: Harry Ruben

VIRUS-MEDIATED TRANSFORMATION
OF MAMMALIAN CELLS

Fred Rapp

Department of Microbiology, College of Medicine
The Milton S. Hershey Medical Center
of The Pennsylvania State University
Hershey, Pennsylvania 17033

Introduction

Animal viruses, both RNA- and DNA-containing, have repeatedly been shown to mediate the conversion of normal cells into cells possessing many new properties, including neoplastic potential. Thus, tranformation of cells by viruses is defined as the acquisition by normal cells following exposure to a virus of new properties which are passed on in a stable heritable form to all progeny cells. The acquisition of some new properties is simple, such as the acquisition by cells unable to produce thymidine kinase of the viral genetic information for the synthesis of this enzyme (Munyon *et al.*, 1971). Other transformational events are more complex, such as the conversion of a normal cell into a cell with oncogenic potential; in this case, many properties of the cell are modified, the cell develops a new transformed phenotype, and many virus functions are expressed. It is this latter type of virus-mediated transformation which will be discussed in detail. Since much information is available for the individual RNA and DNA tumor viruses, the material will be discussed in general terms with specific examples and the reader is referred to one of the many excellent reviews for data on specific viruses (Temin, 1971a; Klein, 1972; Butel *et al.*, 1972; Sambrook, 1972).

General Characteristics of the Transformed Phenotype.

Morphological changes are among the most readily observable modifications of cells following transformation by viruses. Normal cells grow to form a confluent monolayer but contact inhibition then prevents their further growth. Transformed cells no longer respond to contact inhibition and frequently form dense colonies or foci of cells criss-crossing and piling on top of one another (Fig. 1, 2). These cells have other altered growth properties; they attain a higher saturation density, have reduced requirements for specific growth factors, and can grow in soft agar.

The virus genome has been shown to be associated in some stable way with the cell genome. Cell DNA extracted from cells transformed by both DNA and RNA viruses contain all of the genetic information for the replication of these viruses (Boyd and Butel, 1972; Hill and Hillova, 1972). Virus-specific messenger RNA can readily be detected in transformed cells. Virus-specific proteins, such as the SV40 T antigen (Fig. 3), are frequently present.

Cell surface changes are also associated with cells transformed by viruses. Transformed cells are agglutinated more readily by concanavalin A and wheat germ agglutinin than are normal cells. Virus-specified antigens are frequently present on the surface of transformed cells and the unmasking of embryonic antigens at the cell surface also occurs (Fig. 4, 5). All of these virus-induced modifications which form the transformed cell phenotype are summarized in Tables 1 and 2.

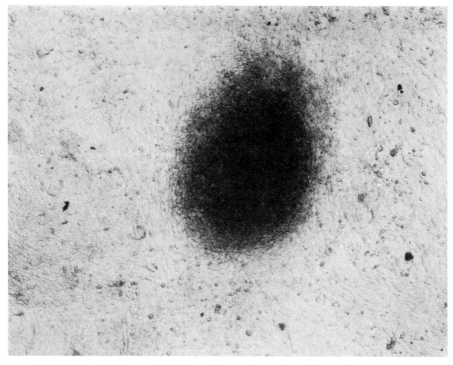

Fig. 1. *A typical focus arising from the transformation of mammalian (hamster) cells by a DNA-containing virus.*

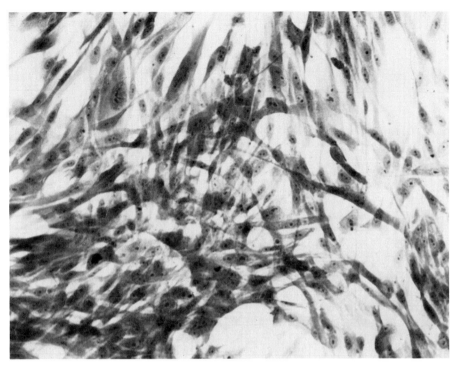

Fig. 2. *Photomicrograph of a transformed culture of hamster cells exhibiting* cross-over pattern.

Ability of conditionally-lethal virus mutants to mediate cell transformation.

Since it has been shown that the virus genome is associated with the cell genome of transformed cells and that virus functions are expressed in the transformed cell, it was of importance to determine the mechanism of virus transformation and the amount of virus genetic information required for transformation. In the beginning, it was possible to postulate that the virus merely de-repressed a function in the cell so that it was irreversibly changed into a cancer cell; under these conditions, the presence of a functioning virus genome was not required for the maintenance of the transformed phenotype. The isolation and utilization of some of the first temperature-sensitive conditional-lethal mutants of viruses known to produce tumors *in vivo* and cell transformation *in vitro* supported this hypothesis. A polyoma virus temperature-sensitive mutant, ts-a, was neither able to replicate nor to transform cells at the non-permissive temperature (Fried, 1965). Therefore, the mutation in this virus mutant involved a virus cistron required for the initiation but not for the maintenance of the transformed state. Similar results

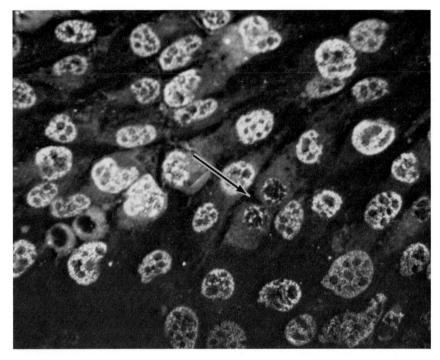

Fig. 3. *Immunofluorescence photomicrograph of SV40 tumor (T) antigen in the nuclei (white) of fixed transformed hamster cells. Arrow points to a cell in mitosis with both daughter cells with T antigen.*

were obtained with a temperature-sensitive mutant of the avian sarcoma viruses (Toyoshima and Vogt, 1969).

However, the characterization of other temperature-sensitive mutants suggested that at least two steps were involved in the transformation of a cell by either DNA or RNA viruses; one step was required for the initiation of the transformation event and the other for the maintenance of the transformed phenotype. Cells transformed at the permissive temperature by polyoma virus mutant ts-3 were shown to lose the transformed phenotype when shifted to the non-permissive temperature (Dulbecco and Eckhart, 1970; Eckhart *et al.*, 1971). Various mutants of Rous sarcoma virus were characterized which could not transform cells at non-permissive temperatures although the virus was capable of replicating under these conditions. Cells transformed at permissive temperatures lost the transformed phenotype when shifted to non-permissive conditions; conversely, cells maintained under non-permissive conditions would demonstrate the transformed phenotype when grown at the permissive temperature (Martin, 1970; Kawai and Hanafusa, 1971; Biquard and Vigier, 1972). Therefore, the virus function of Rous sarcoma virus associated with the maintenance of the transformed phenotype is not required for virus replica-

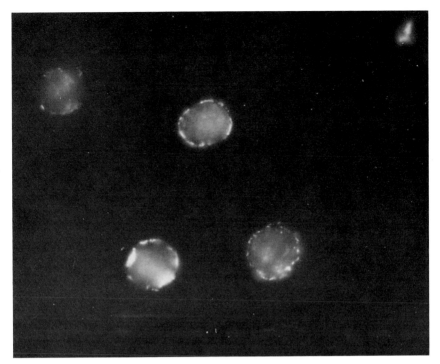

Fig. 4. *Immunofluorescence photomicrograph of viable transformed hamster cells in suspension exhibiting SV40 surface (S) antigen.*

tion; this is in contrast with the information developed with DNA viruses described above.

Complementation tests have also been performed with various of the Rous sarcoma temperature-sensitive mutants. With some of the tests using doubly infected cells, there was an enhancement of the number of foci of transformed cells detected at the non-permissive temperature; no increase in transforming potential was detected when cells were doubly infected with other pairs of mutants (Kawai *et al.*, 1972). These results also suggest that at least two different virus functions are required for the expression of the transformed phenotype. These experiments were not able to differentiate between the possibility of two different proteins or two functional subunits of the same protein.

Studies with host range mutants of polyoma virus also suggest that the virus functions required for the maintenance of the transformed state are continually expressed in the transformed cells. Polyoma virus mutants (such as NG-18) have been isolated which replicate well in transformed cells but which cannot replicate in the non-transformed parental cell line (Benjamin, 1970; Benjamin and Burger, 1970). These mutants are not able to convert normal

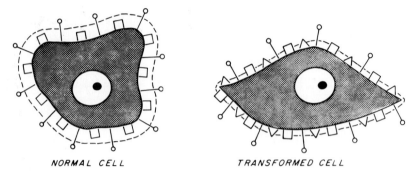

NORMAL CELL   TRANSFORMED CELL

Fig. 5. *Diagrammatic representation of changes at the surface of cells transformed by viruses.*

    □ – *fetal antigens*
    ○ – *normal cell antigens*
    Δ – *virus-specific antigens*

cells into transformed cells. It is suggested that these viruses are able to replicate in the transformed cells because the virus gene functions expressed in the transformed cells provide the gene products which they require for replication; a type of complementation of replication occurs.

Supertransformation studies also suggest that the morphological phenotype of two different DNA viruses can be expressed in the same transformed cell. Cells transformed *in vitro* with polyoma virus were supertransformed by SV40 virus (Todaro *et al.*, 1965). The resulting cells contained virus-specific antigens of each virus. The morphology of cells transformed by each virus independently is so distinct that they are easily separated: polyoma virus transformed cells are long and spindle-shaped and the colonies have a dense piled up center with swirled, star-shaped edges whereas SV40-transformed cells are more plump and epithelioid and the dense colonies have sharp and distinct edges. The supertransformed colonies frequently have elements of each morphological type. The reciprocal experiment, supertransformation of SV40-transformed cells by polyoma virus, has also been done (Takemoto and Habel, 1966).

These supertransformation studies also demonstrate that there is not just one cellular site which must be modified for the expression of cell transformation by DNA viruses; both DNA viruses can express their own transforming potential in the same cell. However, there may be a difference in cell transformation by RNA and DNA viruses. An SV40-transformed cell which does not maintain the transformed phenotype at elevated temperatures can be re-transformed by the RNA murine sarcoma virus at the elevated temperature (Renger, 1972). The temperature-sensitive SV40 transformed cell appears to have a cellular function which is temperature-sensitive because all of the SV40

TABLE 1.
Phenotypic Properties of Transformed Cells

Morphologic Conversion
Increased Saturation Density
Growth in Spent Medium
Growth in Soft Agar
Agglutinable by Plant Lectins
Synthesis of New Antigens
Synthesis of Virus-specific RNA
Synthesis of New Enzymes
Immortality
Oncogenic

virus functions continue to be expressed at the non-permissive temperature and the cells cannot be re-transformed with wild type SV40 virus at the non-permissive temperature; wild type SV40 virus can also be rescued from these cells. The fact that an RNA tumor virus can cause the induction of the transformed phenotype at the non-permissive temperature suggests that either RNA and DNA tumor viruses cause cell transformation by entirely independent pathways or that, if the pathway to eventual cell transformation is common, the RNA tumor viruses must be able to supply a function which the smaller DNA tumor viruses must rely on the cell to supply (the temperature-sensitive cell function in these cells). At the present time, there is no additional evidence to resolve which mechanism is correct.

TABLE 2.
Properties of Cells Transformed by Viruses, Chemicals or Radiation

	Viruses	Chemicals	Radiation
Transforming Agent			
Activation	0	+	0
Toxicity	+ or 0	+	+
Transformed Cells			
Added Genetic Information	+	0	0
New Antigens	+	+	NK
New Nucleic Acids	+	0	0
Common Antigens	+	0	0
Enzymatic Conversion	+	NK	NK
Morphologic Conversion	+	+	+
Increased Saturation Density	+	+	+
Growth in Soft Agar	+	+	NK
Growth in Spent Medium	+	+	NK
Lectin Agglutinability	+	+	+
Immortality	+	+	+
Oncogenic	+ or 0	+ or 0	+ or 0

NK = not known

Ability of defective particles to mediate cell transformation.

Virus particles rendered non-infectious by various external agents have also been shown to mediate cell transformation. Simian papovavirus SV40 inactivated with hydroxylamine was still able to transform cells *in vitro* and to produce tumors *in vivo* (Altstein *et al.*, 1967). Deletion mutants of SV40 virus which could produce the early SV40 T antigen but which could not produce infectious virus were able to transform cells with the same efficiency as virus capable of producing infectious progeny (Uchida and Watanabe, 1969). These results suggest that the entire virus genome is not required to mediate cell transformation. Virus treated with ultraviolet irradiation actually shows an enhanced ability to transform cells (Jensen and Defendi, 1967; Latarjet *et al.*, 1967); the data in Fig. 6 demonstrate that although the ability of PARA-adenovirus 7 to replicate following exposure to ultraviolet irradiation decreases, there is an enhancement of its ability to produce foci following a short exposure to ultraviolet irradiation (Duff *et al.*, 1972). Herpes simplex virus types 1 and 2 and human cytomegalovirus are able to induce cell transformation following exposure to ultraviolet irradiation (Duff and Rapp, 1971a, b; Rapp and Duff, 1972; Duff and Rapp, 1973; Albrecht and Rapp, 1973). Herpes simplex virus types 1 and 2 and SV40 virus inactivated by exposure to fluorescent light following replication in neutral red pretreated cells were also able to effect cell transformation (Rapp *et al.*, 1973). Therefore, only a portion of the virus genome is required to mediate cell transformation and there are suggestions that cell transformations may be more efficient if the ability of a virus to produce infectious progeny and to kill the host cell is reduced by various external agents.

Conclusion

Despite all the changes which have been detected in cells transformed by viruses, there is still no clear idea of the actual mechanisms by which the virus modifies the expression of so many cell properties. No one has yet isolated a "transforming protein" coded for by the virus which of itself will cause the conversion of a normal cell into a transformed cell. There is one postulated mechanism for the maintenance of the transformed state by the DNA viruses SV40 and polyoma (Levine and Burger, 1972). This hypothesis suggests that the DNA viruses code for a protein which causes the initiation of cell DNA synthesis at a place different from the normal initiation site. Thus, cell DNA synthesis is "out of phase." This in turn could modify messenger RNA synthesis, protein synthesis, and normal control mechanisms. However, it is too early to tell if this hypothesis will stand the test of further experimentation.

The above discussion has demonstrated that various RNA and DNA viruses can mediate cell transformation *in vitro*. But, in the final analysis, one

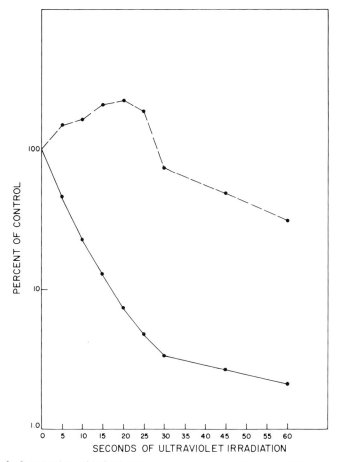

Fig. 6. *Inactivation of infectivity (—•—•—) and transforming ability (—•—•—) of PARA (defective SV40)-adenovirus type 7 by exposure to ultraviolet irradiation. Infectivity is more susceptible to inactivation than is transforming activity. Note that small doses of irradiation enhance transforming activity of the virus.*

is interested in determining if this is the case in natural situations, especially as it concerns man himself. In the rodent system, especially the inbred mice which have proved to be such a useful tool in the laboratory, the prevalence of murine tumor viruses has been so common that two separate theories have been suggested in order to explain the presence of these viruses in normal animals. The virus oncogene hypothesis (Todaro and Huebner, 1972) suggests that all vertebrate cells (both somatic and germ line) contain all of the genetic information required for the production of the type-C RNA tumor viruses. This information is kept in a repressed form in normal cells. However, various external agents, such as radiation, chemicals, and perhaps other viruses, can de-repress this system, allowing the conversion of the normal cell into a transformed cell. The transformed cell may or may not spontaneously release the type C-RNA tumor viruses, depending on the amount of information de-repressed. The protovirus hypothesis (Temin, 1971b) suggests that the germ line cells contain in their chromosomes the potential for genetic evolution by the somatic cells of the information required for cell transformation. Changes leading to cell transformation are not the normal role of this information transfer system but the result in selected cases can be cell transformation.

Originally it was assumed that human cancers were caused by exotic "human tumor viruses." However, evidence is beginning to mount that common viruses, especially herpesviruses, are associated with human cancers (see review by Rapp, 1973). The first association was with human cervical carcinoma and herpes simplex type 2 and with EB virus and Burkitt's lymphoma. More recently, both herpes simplex virus types 1 and 2 and cytomegalovirus have produced cell transformation *in vitro* (Duff and Rapp, 1971a, b; Rapp and Duff, 1972; Duff and Rapp, 1973; Albrecht and Rapp, 1973; Kutinova *et al.*, 1973; Darai and Munk, 1973). Most of these herpesviruses are common in the human population and frequently give rise to persistent or latent infections (see review by Rapp and Jerkofsky, 1973). Recently, human papovaviruses have been isolated from patients suffering from progressive multifocal leucoencephalopathy (Padgett *et al.*, 1971; Weiner *et al.*, 1972) and following renal transplantation (Lecatsas *et al.*, 1973). These isolated viruses resemble the DNA tumor viruses SV40 and polyoma known to be capable of inducing cell transformation. Perhaps more work should be done with common viruses known to produce latent infections in man. Persistence of viral genetic material could easily lead to integration into the host genetic material, interruption of normal genetic pathways, and perhaps neoplastic transformation.

ACKNOWLEDGEMENTS

This study was conducted in part under Contract #70-2024 within the Virus Cancer Program of the National Cancer Institute, NIH, PHS and in part under Public Service Research Grant No. CA11647 from the National Cancer Institute.

Special gratitude is expressed to Dr. Mary Ann Jerkofsky who helped in many ways in the development of this chapter.

REFERENCES

1. Albrecht, T. and Rapp, F. (1973). Malignant transformation of hamster embryo fibroblasts following exposure to ultraviolet-irradiated human cytomegalovirus. *Virology*, 55:53-61.

2. Altstein, A. D., Deichman, G. I., Sarycheva, O. F., Dodonova, N. N., Tsetlin, E. M., and Vassilieva, N. N. (1967). Oncogenic and transforming activity of hydroxylamine-inactivated SV40 virus. *Virology 33*:746-748.

3. Benjamin, T. L. (1970). Host range mutants of polyoma virus. *Proc. Nat. Acad. Sci. USA 67*:394-399.

4. Benjamin, T. L. and Burger, M. M., Absence of a cell membrane alteration function in non-transforming mutants of polyoma virus. *Proc. Nat. Acad. Sci. USA 67*:929-934.

5. Biquard, J. M. and Vigier, P. (1972). Characteristics of a conditional mutant of Rous sarcoma virus defective in ability to transform cells at high temperature. *Virology 47*:444-455.

6. Boyd, V. A. L. and Butel, J. S., (1972). Demonstration of infectious deoxyubonucleic acid in transformed cells. I. Recovery of simian virus 40 from yielder and nonyielder transformed cells. *J. Virol. 10*:399-409.

7. Butel, J. S., Tevethia, S. S. and Melnick, J. L. (1972). Oncogenicity and cell transformation by papovavirus SV40: The role of the viral genome. *Adv. in Cancer Res. 15*:1-55.

8. Darai, G. and Munk, K. (1973). Human embryonic lung cells abortively infected with herpes virus hominis type 2 show some properties of cell transformation. *Nature New Biol. (London) 241*:268-269.

9. Defendi, V. and Jensen, F. (1967). Oncogenicity by DNA tumor viruses: Enhancement after ultraviolet and cobalt-60 radiations. *Science 157:*703-705.

10. Duff, R. and Rapp, F. (1971a). Oncogenic transformation of hamster cells after exposure to herpes simplex virus type 2. *Nature (London) 233*:48-50.

11. Duff, R. and Rapp, F. (1971b). Properties of hamster embryo fibroblasts transformed *in vitro* after exposure to ultraviolet-irradiated herpes simplex virus type 2. *J. Virol. 8*:469-477.

12. Duff, R. and Rapp, F. (1973). Oncogenic transformation of hamster embryo cells after exposure to inactivated herpes simplex virus type 1. *J. Virol. 12:*209-217.

13. Duff, R., Knight, P., and Rapp, F. (1972). Variation in oncogenic and transforming potential of PARA (defective SV40)-adenovirus 7. *Virology 47:*849-853.

14. Dulbecco, R. and Eckhart, W. (1970). Temperature-dependent properties of cells transformed by a thermosensitive mutant of polyoma virus. *Proc. Nat. Acad. Sci. USA 67:*1775-1781.

15. Eckhart, W., Dulbecco, R. and Burger, M. M. (1971). Temperature-dependent surface changes in cells infected or transformed by a thermosensitive mutant of polyoma virus. *Proc. Nat. Acad. Sci. USA 68*:283-286.

16. Fried, M. (1965). Cell-transforming ability of a temperature-sensitive mutant of polyoma virus. *Proc. Nat. Acad. Sci. USA 53*:486-491.

17. Hill, M. and Hillova, J. (1972). Recovery of the temperature-sensitive mutant of Rous sarcoma virus from chicken cells exposed to DNA extracted from hamster cells transformed by the mutant. *Virology 49*:309-313.

18. Kawai, S. and Hanafusa, H. (1971). The effects of reciprocal changes in temperature on the transformed state of cells infected with a Rous sarcoma virus mutant. *Virology 46*:470-479.

19. Kawai, S., Metroka, C. E. and Hanafusa, H. (1972). Complementation of functions required for cell transformation by double infection with RSV mutants. *Virology 49*:302-304.

20. Klein, G. (1972). Herpesviruses and oncogenesis. *Proc. Nat. Acad. Sci. USA 69*:1056-1064.

21. Kutinová, L., Vonka, and Brouček, J., (1973). Increased oncogenicity and synthesis of herpesvirus antigens in hamster cells exposed to herpes simplex type 2 virus. *J. Nat. Cancer Inst. 50:*759-766.

23. Lecatsas, G., Prozesky, O. W., Van Wyk, J. and Els, H. J. (1973). Papova virus in urine after renal transplantation. *Nature (London) 241*:343-344.

22. Latarjet, R., Cramer, R., Golde, A., and Montagnier, L. (1967). Irradiation of oncogenic viruses: Dissociation of viral functions. *In* Carcinogenesis: A broad critique. The Williams and Wilkins Co., Baltimore, Maryland. pp. 677-695.

24. Levine, A. J. and Burger, M. M. (1972). A working hypothesis explaining the maintenance of the transformed state by SV40 and polyoma. *J. Theor. Biol. 37*:435-446.

25. Martin, G. S., (1970). Rous sarcoma virus: A function required for the maintenance of the transformed state. *Nature (London) 227*:1021-1023.

26. Munyon, W., Kraiselburd, E., Davis, D., and Mann, J. (1971). Transfer of thymidine kinase to thymidine kinaseless L cells by infection with ultraviolet-irradiated herpes simplex virus. *J. Virol. 7*:813-820.

27. Padget, B. L., Walker, D. L., ZuRhein, G. M., Eckroade, R. J. and Dessel, B. H. (1971). Cultivation of papova-like virus from human brain with progressive multifocal leucoencephalopathy. *Lancet, i* 1257-1260.

28. Rapp, F. (1973). Question: Do herpesviruses cause cancer? Answer: Of course they do! *J. Nat. Cancer Inst. 50*:825-832.

29. Rapp, F. and Duff, R. (1972). In vitro cell transformation by herpesviruses. *Fed. Proc. 31*;1660-1668.

30. Rapp, F. and Jerkofsky, M. A. (1973). Persistent and Latent Infections, *In* "The Herpesviruses" (ed. A. Kaplan) Academic Press, Inc., New York, New York, pp. 271-289.

31. Rapp, F., Li, J. H., and Jerkofsky, M. (1973). Transformation of mammalian cells by DNA-containing viruses following photodynamic inactivation. *Virology 55:*339-346.

32. Renger, H. C. (1972). Retransformation of ts SV40 transformants by murine sarcoma virus at non-permissive temperature. *Nature New Biol. (London) 240*:19-21.

33. Sambrook, J. (1972). Transformation by polyoma virus and simian virus 40. *Adv. in Cancer Res. 16*:141-180.

34. Takemoto, K. K. and Habel, K. (1966). Hamster tumor cells doubly transformed by SV40 and polyoma viruses. *Virology 30*:20-28.

35. Temin, H. M. (1971a). Mechanism of cell transformation by RNA tumor viruses. *Ann Rev. Micro 25*:609-648.

36. Temin, H. M. (1971b). The protovirus hypothesis: Speculation on the significance of RNA-directed DNA synthesis for normal development and for carcinogenesis. *J. Nat. Cancer Inst. 46*:III-VII.

37. Todaro, G. J. Habel, K., and Green, H. (1965). Antigenic and cultural properties of cells doubly transformed by polyoma virus and SV40. *Virology 27*:179-185.

38. Todaro, G. J. and Heubner, R. J., (1972). The viral oncogene hypothesis: New evidence. *Proc. Nat. Acad. Sci. USA 69*:1009-1015.

39. Toyoshima, K. and Vogt, P. K. (1969). Temperature sensitive mutants of an avian sarcoma virus. *Virology 39*:930-931.

40. Uchida, S. and Watanabe, S. (1969). Transformation of mouse 3T3 cells by T antigen-forming defective SV40 virions (T particles). *Virology 39*:721-728.

41. Weiner, L. P., Herndon, R. M. Narayan, O., Johnson, R. T., Shah, K., Rubinstein, L. J., Preziosi, T. J., and Conley, F. K. (1972). Isolation of virus related to SV40 from patients with progressive multifocal leukoencephalopathy. *New Eng. J. Med. 286*:385-390.

RNA-DIRECTED DNA POLYMERASE ACTIVITY IN UNINFECTED CELLS

Howard M. Temin and Chil-Yong Kang

McArdle Laboratory
University of Wisconsin-Madison

The protovirus hypothesis states that RNA-directed DNA polymerase activity exists in normal cells. The hypothesis has three major corollaries: a) RNA-directed DNA polymerase activity is important in cellular differentiation; b) cellular RNA-directed DNA polymerase activity is the evolutionary precursor of rnadnaviruses; and c) cellular RNA-directed DNA polymerase activity is the site of action of carcinogenic agents (Temin, 1970, 1971, 1972). In this paper, we shall discuss the evidence in favor of the existence of RNA-directed DNA polymerase activity in cells and the evidence bearing on its possible role in normal differentiation.

Earliest reports

The presence of RNA-directed DNA polymerase activity in virions of infectious RNA tumor viruses provided a biochemical model for RNA-directed DNA polymerase activity in normal cells (see Temin and Baltimore, 1972). Coffin and Temin (1971a) looked into Rous sarcoma virus-infected cells for RNA-directed DNA polymerase activity with properties similar to those of virion cores. Such a fraction was found in Rous sarcoma virus-infected chicken cells. Its properties suggested that it might be a virion precursor.

Rous sarcoma virus-infected rat cells were then examined (Coffin and Temin, 1971b). A fraction with endogenous RNA-directed DNA polymerase activity was found. However, experiments with nucleic acid hybridization and with antibodies to avian leukosis-sarcoma virus DNA polymerase showed that neither the RNA template nor the DNA polymerase of this rat cell endogenous RNA-directed DNA polymerase activity was related to the infecting Rous sarcoma virus. This result indicated that the endogenous RNA-directed DNA polymerase activity of RSV-infected rat cells might represent an increase in an activity found in uninfected rat cells.

Experiments carried out with fractions from uninfected rat embryo fibroblasts in culture supported this hypothesis. Low levels of ribo-nuclease-sensitive endogenous DNA polymerase activity was found. Unfortunately, the levels of this activity were so low as to preclude detailed characterization.

Endogenous RNA-directed DNA polymerase Activity in Uninfected Chicken Cells

Experiments were then carried out with normal chicken cells and embryos. Kang and Temin (1972) reported that the microsome fraction from uninfected chicken cells and embryos contained endogenous RNA-directed DNA polymerase activity. This activity was characterized as RNA-directed by its sensitivity to ribonuclease treatment, its resistance to deoxyribonuclease treatment, and its partial resistance to actinomycin D. The product DNA of the endogenous DNA polymerase activity from uninfected chicken embryos annealed to RNA from the same chicken fraction, but not to the RNAs of Rous sarcoma nor reticuloendotheliosis viruses. The chicken endogenous DNA polymerase activity was not neutralized by antibody to avian leukosis-sarcoma virus DNA polymerase.

Further study (Kang and Temin, 1973a) used a modified method of preparation of the chicken activity. With this method, an early DNA product-RNA complex was isolated from a 5-minute reaction of the chicken endogenous RNA-directed DNA polymerase activity in the presence of 100 µg per ml actinomycin D. This complex was characterized by its sedimentation in sucrose gradients and its density in cesium sulfate equilibrium gradients after no treatment or treatments with ribonuclease, Sl nuclease, heat, and alkali. Its structure is shown diagrammatically in Fig. 1.

Treatment with Sl nuclease indicated that the DNA product was about half double-stranded. Furthermore, about the same fraction of the DNA product was resistant to Sl nuclease after heating and quick cooling. Therefore, the DNA rapidly reannealed. This rapidly reannealing indicates that either the DNA product was circular or, more likely, had some self-complementary sequences giving a hairpin structure.

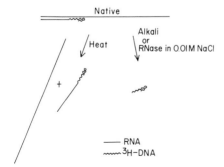

Fig. 1. *Probable structure of early DNA product-RNA complex of chicken endogenous RNA-directed DNA polymerase activity. The product of a five minute reaction in the presence of 100 µg/ml actinomycin D of the chicken endogenous RNA-directed DNA polymerase activity was isolated and characterized.*

Relationship of Uninfected Chicken Cellular Endogenous RNA-directed DNA Polymerase Activity to Avian Rnadnaviruses

There are two groups of avian viruses whose virions contain RNA and a DNA polymerase — the avian leukosis-sarcoma viruses (ALV) and the reticuloendotheliosis viruses (REV). There is some evidence that there can be virus information related to both groups of viruses indigenous to uninfected cells. Therefore, an extensive study was made of the relationship of the chicken endogenous RNA-directed DNA polymerase activity to these avian rnadnaviruses. The DNA product of the chicken endogenous activity made in the presence of actinomycin D did not hybridize to the RNA of several avian leukosis-sarcoma or reticuloendotheliosis viruses. Under the same conditions, 50-100% of the DNA product of the chicken activity annealed to RNA from the same fraction (Table 1). In addition, avian leukosis-sarcoma virus group-specific antigen-negative chicken cells contained endogenous RNA-directed DNA polymerase activity, but did not contain any avian leukosis-sarcoma virus RNA.

The DNA polymerases of the chicken endogenous DNA polymerase activity were studied in two ways. First, the DNA polymerases were isolated from the chicken fraction and were shown to have sedimentation coefficients similar to the soluble 10S and 3-4S DNA polymerases isolated from chicken embryos (Mizutani and Temin, 1973). Some preparations of the 3-4S DNA polymerase were able to transcribe heteropolymer RNA templates, making DNA. The DNA polymerases from avian-leukosis sarcoma viruses and reticuloendotheliosis viruses had larger sedimentation coefficients than this DNA polymerase. Second, studies were made with neutralizing antibodies (Table 2). Antibody which neutralized DNA polymerases of all avian leukosis-sarcoma viruses did not neutralize the chicken endogenous DNA polymerase activity. In addition, antibody made against the chicken 3-4S DNA polymerase neutralized partially or completely (depending on the particular chicken fraction) the endogenous RNA-directed DNA polymerase activity from uninfected chicken cells. This antibody did not neutralize the DNA polymerase activity from either group of avian rnadnaviruses, either in disrupted virions or after partial purification.

Therefore, it is clear that the chicken endogenous RNA-directed DNA polymerase activity is not directly related to either of the avian rnadnaviruses — the avian leukosis-sarcoma viruses or the reticuloendotheliosis viruses.

The Nature of Chicken Endogenous RNA-directed DNA Polymerase Activity

The experiments discussed above provide evidence that endogenous RNA-directed DNA polymerase activity exists in normal uninfected chicken cells and that this activity is not that of any known rnadnavirus. It can be asked whether this cellular activity is related to an unknown rnadnavirus. The

TABLE 1.

Hybridization of the DNA product of chicken endogenous RNA-directed DNA polymerase activity

RNA	DNA product of microsomal fraction isolated from chicken amnion cells		DNA product of microsomal fraction isolated from 4 day chicken embryos	
	$C_r t$ (mole·sec/liter)	% of 270 cpm DNA hybridized	$C_r t$ (mole·sec/liter)	% of 670 cpm DNA hybridized
I-ILV	4.6	0	9.2	0
B77V	–	NT	9.2	3
TDSNV	4.6	0	9.2	0
REV-T	4.6	0	9.2	9
Chicken amnion cells	188	100	376	37
4 day chicken embryo	745	40	1490	53
rat cells	46	0	92	.8
poly(rA)	4.6	0	9.2	0
poly(rI)	4.6	0	9.2	14

[3]H-DNA prepared from the endogenous DNA polymerase activity of microsome fractions was isolated from both ALV-group specific antigen-negative chicken amnion cells and 20 dozen of 4 day mixed chicken embryos. The ^{3}H-DNA was hybridized with 1.25 µg, 12.5 µg, 50 µg, or 200 µg of different RNAs in 50 µl of a hybridization mixture containing 0.3M NaCl, 0.1% of SDS, 0.001 M EDTA, 2% phenol, and 0.05 M Tris-HCl (pH 7.3) at 63°C for 18 or 36 hours. The extent of hybridization was determined by Sl nuclease digestion. 100% radioactivity of the chicken amnion cell fraction DNA was 650 counts/min, and the Sl nuclease resistant fraction in a control incubated without RNA was 380 counts/min. 100% radioactivity of the 4 day chicken embryo fraction DNA was 1570 counts/min, and the Sl nuclease resistant fraction in a control incubated without RNA was 890 counts/min. The background counts were subtracted. (Data from Kang and Temin, 1973c).

NT: not tested.
I-ILV: induced leukosis virus, an ALV.
B77V: strain of avian sarcoma virus.
TDSNV: Trager duck spleen necrosis virus, an REV.
REV-T: reticuloendotheliosis virus strain T.

answer to this question is mainly a matter of definition. If a virus is a genetic entity with a virion which carries the genetic information of the virus from one cell to another, then no virus is present in these normal uninfected chicken cells. However, a less exclusive definition of a virus, which perhaps included episomes, might include something like the endogenous RNA-directed DNA polymerase activity of chicken embryos. For that reason, we have introduced the name protovirus for this activity.

Studies were performed to determine whether this endogenous DNA polymerase activity might have a role in development. One study involved

TABLE 2.
Neutralization of DNA polymerase activities

Antibody to DNA Polymerase	DNA Polymerase			
	ALV[a]	REV[a]	TDSNV[b]	Chicken Fraction[c]
AMV	+	0	±	0
Chicken 10S	0	0	±	0
Chicken 3-4S	0	0	0	± or +

+ Complete neutralization.
± Partial neutralization.
0 No neutralization.
[a] Exogenous DNA polymerase activity of disrupted virions.
[b] Activity of partially purified DNA polymerase.
[c] Endogenous DNA polymerase activity in the presence of actinomycin
       D.

Data from Mizutani and Temin, 1973; Mizutani, Kang, and Temin, 1973.

looking at the relative amount of this activity in different tissues at different times of development of chicken embryos. The results of these studies indicated that this activity was present in different amounts in different tissues and changed in amount during development (Kang and Temin, 1973b). Because of the possible effects of nucleases on the activity, it is difficult to interpret this result.

Secondly, a fraction with endogenous DNA polymerase activity was isolated from different sources: 4 day old chicken embryos, livers of 12 day old chicken embryos, avian leukosis-sarcoma virus group-specific antigen-negative and positive chicken embryo fibroblasts in culture, and nuclei and cytoplasms of avian leukosis-sarcoma virus group-specific antigen-negative chicken amnion cells. The amnion cells were producing budding virions seen with the aid of the electron microscope. (The amnion cells were a kind gift of Dr. R. Dougherty.) In all cases about the same amount of endogenous DNA polymerase activity was observed (Table 3).

To determine if this activity was qualitatively similar in the different tissues, nucleic acid hybridization studies were performed with DNA products of the endogenous activity of pooled 4 day old chicken embryos and chicken amnion cells and RNA from the same fractions (Table 1). In both cases, there was more annealing to RNA from the same fraction and a smaller amount of hybridization to RNA from the other fraction. This result might indicate some common and some tissue-specific sequences in the RNA templates of the endogenous DNA polymerase activity.

Endogenous ribonuclease-sensitive DNA polymerase activity has been reported in other cells, for example, phytohemagglutinin-stimulated human lymphocytes and mouse embryos (Bobrow *et al.*, 1972; Crippa, 1973). However, in no other cases has this activity been carefully studied showing that the product DNA hybridizes to template RNA, isolating an intermediate

TABLE 3.

Endogenous DNA polymerase activity of different chicken cells and tissues

Source of microsome fraction	DNA polymerase activity cpm/100 μg protein/30 min
Pooled 4 day embryos, A	280
Pooled 4 day embryos, B	440
Pooled 12 day embryo livers	300
Cultured fibroblasts, gs positive	220
Cultured fibroblasts, gs negative	220
Cultured amnion cells, gs negative	
nuclei	330
cytoplasms	310

Microsome fractions were prepared, and the endogenous DNA polymerase activity determined in the presence of 100μg/ml actinomycin D (Kang and Temin, 1973a).

gs: ALV group-specific antigen.

RNA-DNA complex, or finding the relationship to rnadnaviruses of the homologous species. In addition, Crippa and Tocchini-Valentini (1971) reported the existence in *Xenopus* oocytes of possible RNA-directed DNA synthesis for the amplification of ribosomal genes. However, a recent report has failed to confirm their findings (Bird *et al.*, 1973). Further experiments must be carried out to clarify whether this activity exists in *Xenopus.*

Other Corollaries of the Protovirus Hypothesis

The hypothesis that cellular endogenous RNA-directed DNA polymerase activity is the evolutionary precursor of rnadnaviruses has little evidence in its favor. There is, however, strong evidence that some rnadnaviruses can be indigenous to cells. In addition, a recent study has shown that the partially purified DNA polymerase of Trager duck spleen necrosis virus, a reticulo-endotheliosis virus, is related to the DNA polymerase of Rous sarcoma viruses and to the chicken large DNA polymerase (Table 2) (Mizutani *et al.*, 1973). (It should be remembered that antibody to the chicken 3-4S DNA polymerase neutralized the chicken endogenous DNA polymerase activity.) This result indicates that there is a possible evolutionary relationship between avian rnadnaviruses and the chicken cellular endogenous RNA-directed DNA polymerase activity.

There is even less evidence that the endogenous RNA-directed DNA polymerase activity of uninfected cells is the site of action of oncogenic agents. No direct experiments have been done to test this hypothesis, however. It remains speculation.

Summary and Conclusions

Endogenous RNA-directed DNA polymerase activity biochemically like that found in virions of rnadnaviruses has been isolated from uninfected chicken embryos and cells. The cellular activity has an RNA template and a DNA polymerase different from that in rnadnaviruses. The DNA polymerase of the cellular endogenous RNA-directed DNA polymerase activity is related to the 3-4S DNA polymerase of cells. There is some evidence for tissue and phase specificity in the amounts of this activity, but its role in differentiation remains hypothetical.

ACKNOWLEDGMENT

This work is supported by Program Project Grant CA-07175 from the National Cancer Institute. C.-Y. Kang is a fellow of the National Cancer Institute of Canada. H. M. Temin holds Research Career Development Award CA-8182 from the National Cancer Institute.

REFERENCES

1. Bird, A., Rogers, E., and Birnstiel, M. (1973). Is gene amplification RNA-directed? *Nature New Biol. 242*, 226-230.

2. Bobrow, S. N., Smith, R. J., Reitz, M. S., and Gallo, R. C. (1972). Stimulated normal human lymphocytes contain a ribonuclease-sensitive DNA polymerase distinct from viral RNA-directed DNA polymerase. *Proc. Nat'l. Acad. Sci, USA 69*, 3228-3232.

3. Coffin, J. M. and Temin, H. M. (1971a). Comparison of Rous sarcoma virus-specific deoxyribonucleic acid polymerases in virions of Rous sarcoma virus and in Rous sarcoma virus-infected chicken cells. *J. Virol., 7*, 625-634.

4. Coffin, J. M. and Temin, H. M. (1971b). Ribonuclease-sensitive deoxyribonucleic acid polymerase activity in uninfected rat cells and rat cells infected with Rous sarcoma virus. *J. Virol., 8*, 630-642.

5. Crippa, M. and Tocchini-Valentini, G. P. (1971). Synthesis of amplified DNA that codes for ribosomal RNA. *Proc. Nat'l. Acad. Sci. USA 68*, 2769-2773.

6. Kang, C.-Y. and Temin, H. M. (1972). Endogenous RNA-directed DNA polymerase activity in uninfected chicken embryos. *Proc. Nat'l. Acad. Sci. USA 69*, 1550-1554.

7. Kang, C.-Y. and Temin, H. M. (1973a). Early DNA-RNA complex from the endogenous RNA-directed DNA polymerase activity of uninfected chicken embryos. *Nature New Biology 242*, 206-208.

8. Kang, C.-Y. and Temin, H. M. (1973b). RNA-directed DNA synthesis in viruses and normal cells: a possible mechanism in differentiation. In: "The Role of RNA in Reproduction and Development," M. Niu and S. Segal (eds.). North-Holland Publ. Co., pp.339-348.

9. Kang, C.-Y. and Temin, H. M. (1973c). Lack of sequence homology among RNAs of avian-leukosis sarcoma viruses, reticuloendotheliosis viruses, and chicken endogenous DNA polymerase activity. *J. Virol. 12*, 1314-1324.

10. Mizutani, S. and Temin, H. M. (1973). Lack of serological relationship among DNA polymerases of avian-leukosis sarcoma viruses, reticuloendotheliosis viruses, and chicken cells. *J. Virol. 12*, 440-448.

11. Mizutani, S., Kang, C.-Y., and Temin, H. M. (1974). Relationships among RNA-directed DNA polymerase activities of avian viruses and chicken cells. *Quant. Biol. 38* (in press).

12. Siu, C-H. and Crippa, M. (1973). RNA-directed DNA polymerase activity in sea-urchin development: specificity of DNA replication. In: "Possible Episomes in Eukaryotes," L. Silvestri (ed.), North-Holland, Amsterdam, pp. 268-275.

13. Temin, H. M. (1970). Malignant transformation of cells by viruses. *Perspectives Biol. and Med., 14*, 11-26.

14. Temin, H. M. (1971). The protovirus hypothesis. *J. Nat'l Cancer Inst., 46*, III-VIII.

15. Temin, H. M. (1972). The protovirus hypothesis and cancer. In: "RNA Viruses and Host Genomes in Oncogenesis," P. Emmelot and P. Bentvelzen (eds.), North-Holland, Amsterdam, London, pp. 351-363.

16. Temin, H. M. and Baltimore, D. (1972). RNA-directed DNA synthesis and RNA tumor viruses. *Adv. Virus Res. 17*, 129-186.

ENDOGENOUS VIRUSES IN NORMAL AND TRANSFORMED CELLS

George J. Todaro, M.D.

National Cancer Institute
Bethesda, Maryland 20014

Over 60 years ago Peyton Rous showed that a virus inoculated into chickens could produce tumors at the site of inoculation by "some unknown mechanism (Rous, 1911)." In the years that have followed, a great many tumor-producing viruses have been found in such animal species as the mouse, the chicken, and the cat, and in these species there is extensive evidence demonstrating that naturally occurring tumors are due to oncogenic viruses (Gross, 1951; Huebner and Todaro, 1969; Synder *et al.*, 1970; Hardy *et al.*, 1973). This probably is the single most important reason for expecting that viruses will prove to be etiologic agents in at least certain human cancers. The active agent that produces tumors in the chicken, in the mouse, and in the cat is an RNA-containing type C virus. While these viruses produce leukemias even under natural conditions (Hardy *et al.*, 1973), it has become clear in the last several years that closely related viruses are found in normal tissues and the genetic information for producing them, in many cases, resides in the cellular DNA. The hypothesis (see Table 1) that the genes for making a complete, potentially infectious virus are present in a repressed form in all somatic cells was first advanced in 1969. (Huebner and Todaro, 1969; Todaro and Huebner, 1972). Endogenous, genetically transmitted type C viruses are represented in the DNA of somatic cells of a great many species in a repressed form. In some tissues they have a greater probability of being expressed than in other tissues. What is not clear is the nature of the relationship between the acquisition of oncogenic potential by a cell and the expression of that cell's endogenous type C viral information. Type C viruses carry oncogenic information and can produce tumors (leukemias, lymphomas, and sarcomas) by exogenous infection; whether the horizontal spread (cell to cell and/or animal to animal) of exogenous type C virus is responsible for a significant proportion of naturally occurring cancers in vertebrates is uncertain; that they *can* have oncogenic potential and *can* produce tumors in a variety of species is firmly established.

TABLE 1.

Some Implications of the Virogene-Oncogene Hypothesis

1. All somatic cells of a species have DNA homologous to type C virus RNA of that species (virogene).

2. Transformed cells should have nucleic acid sequences homologous to the RNA found in transforming viruses.

3. Type C viruses derived from closely related species should have closely related specific antigens, e.g., gs antigens, polymerase and their nucleic acid sequences should be more related to one another than are those viruses released by distantly related species.

While type C viruses have been isolated from tumor tissues of a number of species, they also can be found in normal tissues. Embryonic tissue and placental tissue, in general, show more endogenous type C viral expression than other differentiated tissues (Kalter *et al.*, 1973; Schidlovsky and Ahmed, 1973). In addition, type C viruses can be activated by normal immunologic mechanisms (Hirsh *et al.*, 1972; Armstrong *et al.*, 1972; Armstrong *et al.*, 1973). The viruses produced by both normal tissues and tumorigenic tissues are very similar to one another in their morphological, biochemical, and immunological properties. All of them have reverse transcriptase, an enzyme that is believed to allow the infectious virus to once again become integrated into the host cell DNA (Temin and Kang, this symposium).

With the demonstration by Baltimore (1970) and by Temin and Mitzutani (1970) that RNA tumor viruses contain reverse transcriptase, an enzyme within the virus particle capable of transcribing viral RNA back into DNA, it became important to ask whether this enzyme was specific for oncogenic viruses and whether its presence in cells could be used as a marker for the presence of tumor virus information. A wide variety of viruses that, like the type C RNA tumor viruses, form by budding from the cell membranes have been tested and the results are shown in Table 2. All oncogenic RNA viruses so far tested have this enzyme system, detectable either by an endogenous reaction using the viral RNA or by a polymer-stimulated reaction using synthetic copolymers, such as poly·rA oligo·dT or poly·rG oligo·dC. The synthetic templates increase the sensitivity of enzyme detection, but they are not specific for the oncogenic virus enzyme.

A wide variety of RNA viruses that develop from the cell membrane do not contain the reverse transcriptase. There are, however, some apparent exceptions. The first is visna virus, a virus which causes a chronic, progressive, neurologic disease in sheep that is ultimately fatal. This virus can be recovered from the brains of both diseased and normal sheep, and, unlike most tumor viruses, is cytopathic for cells in tissue culture. However, there is evidence from studies by Takemoto and Stone (1971) that the virus can transform mouse cells and that the transformed cells will produce tumors when

TABLE 2.
DNA Polymerase in RNA-containing Viruses

Virus	Endogenous	Polymer-Stimulated
Oncogenic		
Type C (11 species)	+	+
Type B (MTV*)	+	+
Non-oncogenic		
Sendai	−	−
Influenza	−	−
Respiratory syncytial	−	−
Newcastle disease	−	−
Vesicular stomatitis	−	−
Lymphocytic choriomeningitis	−	−
Possibly oncogenic		
Visna	+	+
Foamy (4 species)	+	+
Monkey mammary	+	+

*Mouse mammary tumor virus.

inoculated in mice. So it may well be that this, too, is a virus that does have oncogenic potential. The second major exceptions are the group of "foamy" viruses. These RNA-containing viruses are frequently found in healthy as well as diseased monkeys, cattle, and cats; they have not yet been associated with any disease. Visna and foamy viruses, then, are apparent exceptions to the rule that only tumor viruses contain RNA-directed DNA polymerase

The foamy viruses of monkeys, cats, and cattle are an interesting group of reverse transcriptase-containing viruses (Parks *et al.*, 1971) and are very common in these three species. Foamy viruses are capable of remaining latent in animals for long periods without overt expression of disease. Whether these viruses are capable of producing *in vitro* cell transformation is as yet unclear. Their typical effect is to alter the membranes of cells that they infect so that the cell membranes will fuse with the membranes of adjacent cells to form syncytia. The viruses can be titered for biological activity by their syncytium-forming ability. Their biologic properties are very much like those of the oncogenic viruses; replication is blocked by inhibitors such as BrdU and by actinomycin D (Parks and Todaro, 1972). The expression of the transformed state, in this case an altered cell membrane that will fuse with other cells, again requires cell division, as does transformation by oncogenic viruses. There is no evidence that the viruses in fact are oncogenic; however, this point has not been adequately tested.

The mammalian foamy viruses may have their counterparts in the avian system in the reticuloendotheliosis group, in that they are readily distinguished from the type C viruses, are genetically cytopathic, and readily spread under normal conditions from animal to animal as a more typical horizontally transmitted virus infection (Purchase *et al.*, 1973; Kang and Temin, 1973).

So, two apparent exceptions (visna and foamy viruses) may really turn out not to be exceptions. They both may be viruses that, like the known tumor viruses, are, under certain conditions, capable of transforming cells. In this case, the reverse transcriptase would have been very valuable for it would allow us to detect viruses that have oncogenic potential. An alternate explanation, however, is that RNA-containing viruses that have reverse transcriptase have the capacity to form a stable association with the infected cell presumably becoming integrated into the cellular genome, and that this property is not necessarily linked to oncogenic capacity. The class of viruses that have reverse transcriptase, then, may be considerably larger than the class of viruses that also have oncogenic potential. Since the foamy viruses are so prevalent in nature and since they can be transmitted horizontally, the finding of reverse transcriptase in a virus does not necessarily prove that the virus is oncogenic.

Type C viruses can be detected in a variety of ways (Table 3). For example, by electron microscopy typical type C viruses can be visualized (Fig. 1). There is some difficulty, however, in sampling large enough areas; consequently, the method is not particularly sensitive for detecting low levels of virus production. In addition, there is some difficulty in distinguishing type C viruses from closely related viruses. The type C viruses can also be recognized by their biologic activity — the ability to form foci or to transmit sarcoma virus from a nonproducer transformed sarcoma cell to another cell. To demonstrate biologic activity requires a cell susceptible to infection by the virus being tested. As a general rule (see below), the cell from which the virus emerges very often is resistant to exogenous infection by the same virus.

All mammalian type C viruses have a common interspecies group-specific antigen which can be detected by complement fixation tests or, with much

TABLE 3.
Detection of Type C Viruses in Cell Cultures and Tissues

1. *Electron microscopy*: tumors and tissue culture often difficult to distinguish from other, related viruses and from cellular particles.

2. *Biologic activity*: focus formation, "rescue" of nonproducer, transmissible biologic activity specific interference with related infectious virus.

3. *Interspecies group-specific antigens of type C viruses*: immunodiffusion, complement fixation, radioimmunoassay.

4. *Virus-specific reverse transcriptase*: immunoadsorbant column, polymerase antibody.

5. *Genetic information*: cellular nucleic acid hybridizable to viral RNA or to the DNA produced *in vitro* from type C viral RNA.

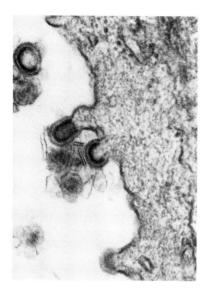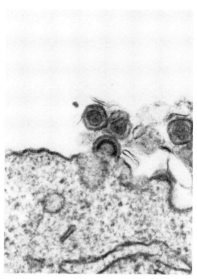

Fig. 1. *Electron micrographs of budding type C viruses from IdU-treated cultures (30 μg/ml for 24 hours). The cells were fixed three days after IdU exposure. On the left –* SV3T3, *clone 5; on the right –* Balb/3T3, *clone S2. x100,000. (Reproduced from Todaro,* Membranes and Viruses in Immunopathology, *1972, by permission of Academic Press, Inc.)*

greater sensitivity, by radioimmunoassay (Parks and Scolnick, 1972; Parks *et al.*, 1973; Oroszlan *et al.*, 1972). Similarly, all mammalian type C viruses have a reverse transcriptase that has interspecies properties (Scolnick *et al.*, 1972). Antiserum made to the polymerase of one mammalian type C virus not only efficiently inhibits the enzymatic activity of the homologous viral polymerase, but will also neutralize other mammalian type C viral polymerases without affecting the normal cellular DNA polymerases (Ross *et al.*, 1971). The reverse transcriptase assay, especially when synthetic templates are used, is a very sensitive assay for detecting low levels of virus released from cell cultures into the supernatant fluid. A fifth approach is to search for viral specific genetic information either in the cells themselves (Baluda and Nayak, 1970; Varmus *et al.*, 1972b) or in particles that have properties similar to those of known viruses (Baxt and Spiegelman, 1972). The molecular hybridization method has been extensively exploited because it permits the detection of partial expression of viral genetic information. It allows the potential recognition of a class of cells that have some expression of viral specific genetic information in them, but not enough to make a complete virus (Tsuchida *et al.*, 1971; Parks *et al.*, 1973).

Balb/c mouse embryo cells give rise to continuous cell lines. These cell lines spontaneously transform and acquire the ability to produce tumors (Aaronson and Todaro, 1968). Some of the lines that are spontaneously transformed also begin producing type C viruses, and the spontaneous production of type C viruses can be demonstrated from single cell clones that have spontaneously transformed (Todaro, 1972; Lieber and Todaro, 1973). Once a subclone spontaneously starts to make virus, it continues to produce that virus indefinitely (Todaro, 1972) (Fig. 2). There must be a mechanism that allows control of the endogenous viral information, and this mechanism can be disturbed in such a way that continued production of the virus results in cells that are nevertheless able to proliferate. The transformed cells are recognized by their loss of contact inhibition, their ability to grow over one another in a random fashion to form multiple cell layers (Fig. 3) and their ability to produce tumors upon inoculation into a susceptible host.

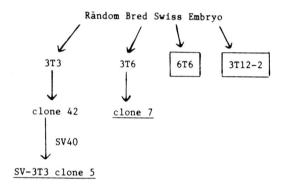

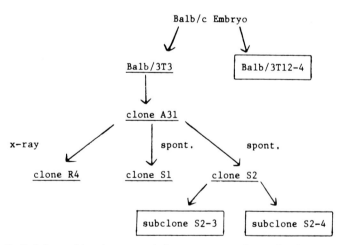

Fig. 2. *Cell lines with a box around them represent cultures that have spontaneously released type C virus; those underlined have endogenous type C virus detectable by induction with iodeoxyuridine (IdU, 30 μg/ml for 24 hours). Subclones S2-3 and S2-4 have been through more than 300 cell generations and more than 2 years in culture before spontaneously releasing virus. (Reproduced from Todaro,* Nature New Biology 240, 157, *1972, by permission of Macmillan and Co.)*

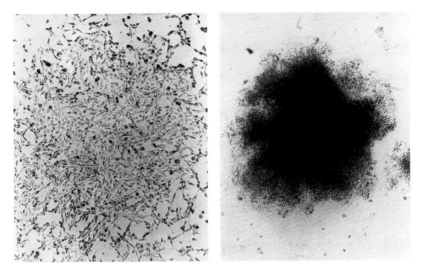

Fig. 3. *On the left is a colony of Balb/c 3T3 clone A31 fixed 10 days after inoculating a single cell onto a Petri dish. On the right is a colony of S2, a spontaneously transformed subclone derived from A31. The individual cells in the S2 colony are epithelioid and tend to readily grow over one another. The transformed colony is considerably denser and more compact. (Reproduced from Todaro,* Nature New Biology 240, *157, 1972, by permission of Macmillan and Co.)*

While certain of the transformed cells spontaneously produce virus, others do not. However, of the ones that do not, there is a class which is superinducible for virus production; these cells will respond much more rapidly and yield much higher titers of virus production when treated with inducers, such as 5-bromodeoxyuridine (BrdU) (Fig. 4). Virus can be detected within the first 8 hours after the addition of the inducer and continues a logarithmic increase for the first two days (Lieber, *et al.*, 1973a) after which time it plateaus and then either slowly declines or continues to secrete virus indefinitely. With the untransformed cells, virus production is slower and shuts off more rapidly (Todaro, 1972; Lieber and Todaro, 1973; Todaro and Huebner, 1972). With the diploid cell strains, production of virus is even less on a per cell basis and lasts for a shorter time. The thymidine analogs can disturb the normal control mechanism, and the effect of this perturbation is most extreme in the most transformed cells. The diploid cells recover quickly; the aneuploid, but still nontumorigenic, cells respond more rapidly but turn off virus production. The transformed cells respond most rapidly and may or may not then turn off virus production. Clearly, the genetic information for production of a complete type C virus is present in a form that allows it to be rapidly called for when the proper inducers are added. There are as yet no examples in which virus has been induced by chemical agents where it has not also been found to appear spontaneously in the same cell lines; the thymidine

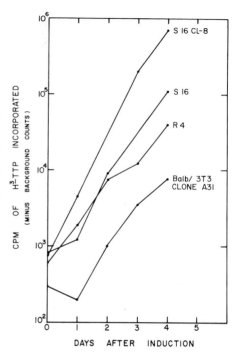

Fig. 4. *BrdU induction of Balb/3T3 and its transformed derivatives R4, S16, and S16 Cl-8; 33 μg/ml of BrdU were applied to just confluent cultures of the cells in 250-ml plastic tissue culture flasks for 24 hours. The 15 ml of medium were changed daily and assayed for reverse transcriptase activity. (Reproduced from Lieber and Todaro,* Int. J. Cancer 11, *616, 1973.)*

analogs increase the probability of an event (secretion of type C virus) that nevertheless occurs spontaneously. Recent evidence suggests that there are at least three recognizably different type C viruses in Balb/3T3 cells that differ from one another both in their host range and in their molecular hybridization (Benveniste *et al.*, 1974).

There are now at least seven species in which it can clearly be shown that a complete copy of the endogenous type C virus, the virogene (Huebner and Todaro, 1969; Todaro and Huebner, 1972), is present in the normal cells (Table 4). Certain continuous lines of the Chinese hamster, mouse, chicken, rat (Duc-Nugyen *et al.*, 1966), and cat all have single cell clones that have the potential to respond rapidly to BrdU and in so doing produce a complete type C virus. One group of type C viruses induced from mouse cell cultures grows in primate and other nonrodent cells, but is unable to infect mouse cells. The endogenous virus from cat cells grows readily in primate, including human, cells but not in most cat cell strains (Todaro *et al.*, 1973). The endogenous Chinese hamster, rat and pig type C viruses (see Table 4) have so far neither been transmissible to cells of other species nor infectious for their own cells (Lieber *et al.*, 1973b; Todaro *et al.*, 1974). One essential attribute, then, of endogenous viruses would appear to be this inability to reinfect the cell lines from which they emerge. This would suggest that it may well be easier to

TABLE 4.

Species Where a Complete Virogene is Known to be Present in Normal Cells

Chicken
Chinese hamster*
Syrian hamster
Mouse*
Rat*
Cat*
Pig*

*Single cell clones spontaneously produce virus in long term culture.

isolate the endogenous virus of a species using the cells of an unrelated species as the permissive host. There now exist several cell lines that secrete endogenous virus but are not reinfected by it. These represent the only examples in which study of a pure endogenous virus is possible, because with the infectious viruses that are transmitted horizontally, i.e., from cell to cell, there is the very real risk that the virus produced will be a mixture, either phenotypic or genetic, of the infecting virus and the endogenous virus of the cell in which the infecting virus is grown. The more a virus is passed from cell to cell the more likely that the virus produced is, in fact, a mixture of several genetically different viruses.

Properties of endogenous viruses include their inability to infect cells that produce them, their presence in many (perhaps all) species, and their greater likelihood to be produced by transformed rather than by normal cells. The transformed cells that produce virus appear to be less tumorigenic than the transformed cells that do not produce virus (Barbieri *et al.*, 1971; Lieber and Todaro, 1973). This, coupled with the fact that transformed cells are much more likely to release endogenous virus, suggests that such a system might have a protective role: virus expression might protect on an immunologic basis against tumor formation. A system that would turn on virogene information when the oncogenic information is already expressed would be of great benefit to the host. This leads to the possibility that the endogenous virus under natural circumstances may serve more to protect against cancer than act as an etiologic agent. The production of the new virus would result in a variety of antigenic changes in the cells that in an immunologically competent animal might facilitate the rejection of those cells. One of the important biologic mysteries is the fact that the system which allows a complete type C virus to be rapidly called forth is present in so many species and is preserved through hundreds of generations in cell culture. For such a system to have been preserved in the evolution of the species, it must have had, on balance, a selective advantage for the host. One possibility would be that virus production serves as an early warning system to the animal that there are cells in the body that have already become transformed. However, this would not explain why the system is preserved for hundreds of cell generations in culture. Over 100 independently isolated subclones of Balb/3T3 clone A31 have now been tested, and every one of them has inducible type C virus. Prior to chemical induction or spontaneous activation there is partial, but not complete, expression of the viral type C virus-specific information (Parks *et al.*, 1973).

The system involving the activation of the type C virus has obvious superficial similarities to the lysogenic system in bacteria. However, in many ways virogene induction might be considered more analogous to the switching on of a differentiated function by vertebrate cells (Stellwagen and Tomkins, 1971). Either spontaneously or after addition of a small molecule, the cell begins producing proteins whose genes are normally repressed and assembles a complex package for export from the cell. BrdU has been known to greatly affect the differentiated state in culture. Upon induction with thymidine analogues, new viral RNA rapidly appears. One major control in this system, then, is at the level of transcription. Whether there will be additional controls at the level of translation remains to be resolved. It is possible that the system works entirely by reading off cellular genes. The cells lines that produce endogenous virus would not actually be replicating the virus, but would, rather, be transcribing and translating information that is part of their natural genetic makeup.

These cells, then, would be expressing a specific differentiated function, e.g., the production of virus. The control of mammary tumor virus (MTV) expression in the mouse may be similar. MTV genomes are also present in all mouse tissues (Varmus, *et al.*, 1972a); however, expression of virus-specific information is restricted to breast tissue and to tumor cells derived from breast tissue. The finding of completed type C viral expression in normal mouse embryo cells at the blastocyte stage – but not earlier (Chase and Pikó, 1973) – and not in most of the adult tissues, is constant with the concept that viral expression can be considered a differentiated function; all cells have the capability but all cells do not continuously produce these endogenous viruses. Different levels of viral expression have been found in different tissues even from the same animal (Vaheri and Ruoslahti, 1973; Parks *et al.*, 1973). Why such a system has been so well preserved in evolution is not clear. The possible bases for the persistence of this system might be:

1. To provide resistance to related, but more virulent viruses. The related virulent viruses could be more lethal or more oncogenic than the resident endogenous virus. Cells having endogenous type C viruses might be protected against infection by the disease producing viruses.

2. The system could have a selective advantage to the host on an immunologic basis if activation of oncogenic information by cells is associated with type C virus release as it seems to be for spontaneously transformed Balb/3T3 cells. In this case the production of virus by transformed cells could

serve as an immunologic signal that the host's immune system could then respond to. Type C virus production, then, in normal circumstances could serve actually to decrease the potential for spontaneous tumor development by facilitating the immune defense system.

3. The virus or certain viral components could serve some normal function during development either of the embryo or of the trophoblast. It need only perform this essential function for a brief time during development or perhaps only in a specific organ for the system to be preserved. Developing cells grow rapidly, and cells such as tropholast cells can be highly invasive, having many of the properties associated with tumor cells. Expression of the virus-related oncogenic information, though inappropriate in the adult, may be part of the normal developmental process.

4. The virus is a highly effective intracellular parasite. It is linked to some critical cell functions that can not be deleted, functions, for example, that are concerned with the initiation or regulation of cell division.

With the virus systems that have been adequately studied, the evidence indicates that there are multiple copies (8 to 12 per diploid genome) and some species such as the mouse have related, but different, viral genomes (virogenes) in their DNA (Aoki and Todaro, 1973; Aaronson and Stephenson, 1973). The endogenous viruses activated by chemical means or appearing spontaneously clearly do not have the oncogenic potential of highly virulent laboratory strains of type C viruses. Those viruses most commonly used in cancer research are type C viruses that have been selected for their oncogenic capacity. There is not enough data in enough species yet to determine the oncogenic potential of the various naturally occurring endogenous type C viruses.

Genetic experiments between high and low leukemic mouse strains demonstrate that type C virus must be expressed and not merely in the latent form in order to produce disease (Meier *et al.*, 1973). The greater the level of type C virus expression the greater the likelihood that the animal will develop tumors. This suggests that there is, in fact, a link between endogenous type C virus expression and subsequent risk of cancer.

REFERENCES

1. Aaronson, S. A., and Stephenson, J. R. (1973). Independent segregation of loci for activation of biologically distinguishable RNA C-type viruses in mouse cells. *Proc. Natl. Acad. Sci. USA 70*, 2055-2058.

2. Aaronson, S. A., and Todaro, G. J. (1968). Development of 3T3-like lines from Balb/c mouse embryo cultures: Transformation susceptibility to SV40. *J. Cell Physiol. 72*, 141-148.

3. Aoki, T., and Todaro, G. J. (1973). Antigenic properties of endogenous type C viruses from spontaneously transformed clones of Balb/3T3. *Proc. Natl. Acad. Sci. USA 70*, 1598-1602.

4. Armstrong, M. Y. K., Black, F. L., and Richards, F. F. (1972). Tumor induction by cell-free extracts derived from mice with graft versus host disease. *Nature New Biol.* 235, 153-154.

5. Armstrong, M. Y. K., Ruddle, N. H., Lipman, M. B., and Richards, F. F. (1973). Tumor induction by immunologically activated murine leukemia virus. *J. Exp. Med.* 137, 1163-1179.

6. Baltimore, D. (1970). Viral RNA-dependent DNA polymerase. *Nature* 226, 1209-1211.

7. Baluda, M. A., and Nayak, D. P. (1970). DNA complementary to viral RNA in leukemic cells induced by avian myeloblastosis virus. *Proc. Natl. Acad. Sci. USA* 66, 329-336.

8. Barbieri, D., Belehradek, J. Jr., and Barski, G. (1971). Decrease in tumor-producing capacity of mouse cells following infection with mouse leukemic virus. *Int. J. Cancer* 7, 364-371.

9. Baxt, W. G., and Spiegelman, S. (1972). Nuclear DNA sequences present in human leukemic cells and absent in normal leukocytes. *Proc. Natl. Acad. Sci. USA* 69, 3737-3741.

10. Benveniste, R. E., Lieber, M. M., and Todaro, G. J. (1974). A distinct class of inducible murine type C viruses which replicate in the rabbit SIRC cell line. *Proc. Natl. Acad. Sci. USA* 71, 602-606.

11. Chase, D. G., and Pikó, L. (1973). Expression of A- and C-type particles in early mouse embryos. *J. Natl. Cancer Inst.* 51, 1971-1975.

12. Duc-Nugyen, H., Rosenblum, E. N., and Zeigel, R. D. (1966). Persistent infection of a rat kidney cell line with Rauscher murine leukemia virus. *J. Bacteriol.* 92, 1133-1140.

13. Gross, L. (1951). "Spontaneous" leukemia developing in C3A mice following inoculation, in infancy, with Ak-leukemic extracts, or Ak-embryos. *Proc. Soc. Exp. Biol. Med.* 76, 27.

14. Hardy, W. D., Jr., Old, L. J., Hess, P. W., Essex, M., and Cotter, S. (1973). Horizontal transmission of feline leukaemia virus. *Nature* 244, 266-269.

15. Hirsch, M. S., Phillips, S. M., Solnik, C., Black, P. H., Schwartz, R. S., and Carpenter, C. B. (1972). Activation of leukemia viruses by graft-versus-host and mixed lymphocyte reactions *in vitro. Proc. Natl. Acad. Sci. USA* 69, 1060-1072.

16. Huebner, R., and Todaro, G. J. (1969). Oncogenes of RNA tumor viruses – as determinants of cancer. *Proc. Natl. Acad. Sci. USA* 64, 1087-1094.

17. Kalter, S. S., Helmke, R. J., Panigel, M., Heberling, R. L., Felsburg, P. J., and Axelrod, L. R. (1973). Observations of apparent C-type particles in baboon (Papio cynocephalus) placentas. *Science* 179, 1332-1333.

18. Kang, C-Y., and Temin, H. M. (1973). Lack of sequence homology among RNAs of avian leukosis-sarcoma viruses, reticuloendotheliosis viruses, and chicken endogenous RNA-directed DNA polymerase activity. *J. Virol.* 12, 1314-1324.

19. Lieber, M. M., and Todaro, G. J. (1973). Spontaneous and induced production of endogenous type C RNA virus from a clonal line of spontaneously transformed Balb/3T3. *Int. J. Cancer* 11, 616-627.

20. Lieber, M. M., Livingston, D. M., and Todaro, G. J. (1973a). Superinduction of endogenous type C virus by 5-bromodeoxyuridine from transformed mouse clones. *Science* 181, 443-444.

21. Lieber, M. M., Benveniste, R. E., Livingston, D. M., and Todaro, G. J. (1973b). Mammalian cells in culture frequently release type C viruses. *Science* 182, 56-59.

22. Meier, H., Taylor, B. A., Cherry, M., and Huebner, R. J. (1973). Host-gene control of type-C RNA tumor virus expression and tumorigenesis in inbred mice. *Proc. Natl. Acad. Sci. USA* 70, 1450-1455.

23. Oroszlan, S., White, M. M. H., Gilden, R. V., and Charman, H. (1972). A rapid direct radioimmunoassay for type C virus group-specific antigen and antibody. *Virology* 50, 294-296.

24. Parks, W. P., and Scolnick, E. M. (1972). Radioimmunoassay of mammalian type C viral proteins: Interspecies antigenic reactivities of the major internal polypeptides. *Proc. Natl. Acad. Sci. USA* 69, 1766-1770.

25. Parks, W. P., and Todaro, G. J. (1972). Biologic properties of syncytium-forming ("foamy") viruses. *Virology* 47, 673-683.

26. Parks, W. P., Livingston, D. M., Todaro, G. J., Benveniste, R. E., and Scolnick, E. M. (1973). Radioimmunoassay of mammalian type C viral proteins. III. Detection of viral antigen in normal murine cells and tissues. *J. Exp. Med.* 137, 622-635.

27. Parks, W. P., Scolnick, E. M., Todaro, G. J., and Aaronson, S. A. (1971). RNA-dependent DNA polymerase in primate syncytium-forming ("foamy") viruses. *Nature* 229, 257-260.

28. Purchase, H. G., Ludford, C., Nazerian, K., and Cox, H. W. (1973). A new group of oncogenic viruses: Reticuloendotheliosis, chick syncytial, duck infectious anemia, and spleen necrosis viruses. *J. Natl. Cancer Inst.* 51, 489-499.

29. Ross, J., Scolnick, E. M., Todaro, G. J., and Aaronson, S. A. (1971). Separation of murine cellular and murine leukemia virus DNA polymerase. *Nature New Biol.* 231, 163-167.

30. Rous, P. (1911). A sarcoma of the fowl transmissible by an agent separate from tumor cells. *J. Exp. Med.* 13, 397.

31. Schidlovsky, G., and Ahmed, M. (1973). C-type virus particles in placentas and fetal tissues of rhesus monkeys. *J. Natl. Cancer Inst.* 51, 225-233.

32. Scolnick, E. M., Parks, W. P., and Todaro, G. J. (1972). Reverse transcriptases of primate viruses as immunological markers. *Science* 177, 1119-1121.

33. Snyder, S. P., Theilen, G. H., and Richards, W. P. C. (1970). Morphological studies on transmissible feline fibrosarcoma. *Cancer Res.* 30, 1658-1667.

34. Stellwagen, R. H., and Tomkins, G. M. (1971). Differential effect of 5-bromodeoxyuridine on the concentrations of specific enzymes in hepatoma cells in culture. *Proc. Natl. Acad. Sci. USA* 68, 1147-1150.

35. Takemoto, K. K., and Stone, L. B. (1971). Transformation of murine cells by two "slow viruses," visna virus and progressive pneumonia virus. *J. Virol.* 7, 770-775.

36. Temin, H. M., and Kang, C-Y. (1974). RNA-directed DNA polymerase activity in uninfected cells. *In* "Developmental Aspects of Carcinogenesis and Immunity" (T. J. King, ed.), Academic Press, New York.

37. Temin, H. M., and Mizutani, S. (1970). RNA-dependent DNA polymerase in virions of Rous sarcoma virus. *Nature* 226, 1211-1213.

38. Todaro, G. J. (1972). "Spontaneous" release of type C viruses from clonal lines of "spontaneously" transformed Balb/3T3 cells. *Nature New Biology* 240, 157-160.

39. Todaro, G. J., and Huebner, R. J. (1972). The viral oncogene hypothesis: New evidence. *Proc. Natl. Acad. Sci. USA* 69, 1009-1015.

40. Todaro, G. J., Benveniste, R. E., Lieber, M. M., and Livingston, D. M. (1973). Infectious type C viruses released by normal cat embryo cells. *Virology* 55, 506-515.

41. Todaro, G. J., Benveniste, R. E., Lieber, M. M., and Sherr, G. J. (1974). Characterization of a type C virus released from the porcine cell line (PK(15). *Virology* 58 65-74.

42. Tsuchida, N., Robin, M. S., and Green, M. (1971). Viral RNA subunits in cells transformed by RNA tumor viruses. *Science* 176, 1418-1419.

43. Vaheri, A., and Ruoslahti, E. (1973). Expression of the major group specific protein (GS-A) of avian type C viruses in normal chicken cells and tissues. *Int. J. Cancer* 12, 361-367.

44. Varmus, H. E., Bishop, J. M., Nowinski, R. C. and Sarker, N. H. (1972a). Mammary tumour virus specific nucleotide sequences in mouse DNA. *Nature New Biol.* *238*, 189-190.

45. Varmus, H. E., Weiss, R. A., Friis, R. R., Levinson, W., and Bishop, J. M. (1972b). Detection of avian tumor virus-specific nucleotide sequences in avain cell DNAs. *Proc. Natl. Acad. Sci. USA* *69*, 20-24.

SESSION V
Immunity and Oncogenesis

Moderators: Norman G. Anderson
Richmond T. Prehn

DEVELOPMENT AND DIFFERENTIATION
OF LYMPHOCYTES

Martin C. Raff

Medical Research Council Neuroimmunology Project,
Zoology Department
University College London
London WC1E 6BT

Lymphocytes are at the center of the immunological universe, in that they are the cells that respond specifically to antigenic challenge. Working in collaboration with other, non-lymphoid cells, such as macrophages, granulocytes, and mast cells, lymphocytes defend the body against infection and tumor overgrowth. The development of immunocompetent lymphocytes and their differentiation into activated effector cells on encounter with specific antigen, provides a complex, yet readily accessible model of differentiation. In this paper, I will briefly review the general properties of lymphocytes and their roles in immunity, and consider in more detail their origins and sequential differentiation.

LYMPHOCYTES AND THE CELLULAR BASIS OF IMMUNITY

It has long been known that there are two general types of immune response, those that can be adoptively transferred to another animal by means of cells, termed *cell-mediated immunity* and those that can be transferred by serum antibody, called *humoral (antibody) immune responses.* Only in the last five or ten years has it been appreciated that these two different types of response are mediated by two distinct classes of lymphocytes, now referred to as T and B lymphocytes respectively. The terms T and B refer to the origins of these cells, T lymphocytes developing from cells in the thymus gland and B lymphocytes from cells in the bursa of Fabricius in birds or from the equivalent tissue (not yet identified) in mammals. Within the thymus and bursa (or bursa equivalent) circulating hemopoietic stem cells differentiate to lymphocytes which, following further maturation, migrate out to the peripheral lymphoid tissues (e.g. spleen, lymph nodes, gut-associated lymphoid tissues) where they make up the populations of immunocompetent T and B lymphocytes respectively (Owen, 1972).

Perhaps the most characteristic feature of immunocompetent T and B lymphocytes is their ability to recognize and respond to antigens in a remarkably specific manner. As predicted by the clonal selection hypothesis (Burnet, 1959) it has now been demonstrated that at some time in their development individual T and B cells become committed to responding to one or a relatively small number of antigens, and express this commitment by displaying antigen-specific receptors on their surface. Thus, when an antigen enters the body it selects out those lymphocytes which already have receptors for the antigen; the interaction of antigen with receptors initiates the activation of the specific cells, leading to the proliferation and/or differentiation of the resting lymphocytes to effector cells. In the case of B cells, the receptors are immunoglobulin (antibody) molecules (Singhal and Wigzell, 1971), and activation induces the lymphocytes to become antibody-secreting cells, the secreted antibody having the same specificity as the surface receptor on the parent B cell (Cozenza and Köhler, 1972). The chemical nature of the T cell receptor is still in dispute, with some controversial evidence for Ig receptors co-existing with evidence for non-Ig receptors (Crone et al., 1972). The most likely candidates for non-Ig T cell receptors are the products of the Immune Response (IR) genes that are genetically linked to the major histocompatibility loci and which influence T cell responses to a variety of antigens (Shevach et al., 1972). When T cells are activated by antigen they secrete a variety of factors ("lymphokines") which are not antigen-specific, such as macrophage migration inhibition factors (MIF), chemotactic factors, cytotoxic factors and mitogenic factors, at lease some of which presumably play a role in cell-mediated immune responses for which T cells are primarily responsible (Lawrence and Landy, 1959). These responses include delayed hypersensitivity, contact sensitivity, rejection of foreign or abnormal tissues, graft-versus-host responses and immunity to some microbes, and in all of them, T cells enlist the help of macrophages. In addition, under some circumstances, activated T lymphocytes can kill "foreign" target cells directly (Brunner and Cerottini, 1971).

Thus, activated B cells secrete antibody and are responsible for humoral antibody responses, while activated T cells secrete lymphokines and are responsible for cell-mediated immunity, However, the dichotomy is not absolute and the immune system not that simple. It is now known that antibody produced by B cells can enhance or inhibit the response of other B cells or T cells responding to the same antigen, and T cells can enhance or inhibit the response of other T cells or B cells. For example, an important recent discovery in immunology has been the demonstration that T cells play an important role in helping B cells to make antibody responses to most immunogens. Although the exact mechanism involved in this T-B cell cooperation is still unclear, it has been shown that it involves an "antigen-bridge" between T cell receptors recognizing one antigenic determinant on an immunogen and B cell receptors recognizing a different determinant on the same immunogen (Mitchison *et al.*, 1970).

Besides differing in their origins and functions, T and B lymphocytes differ in almost every property that has been studied, including their location in peripheral lymphoid tissues, extent of recirculation between blood and lymph, surface charge, adherance to various materials, mitogenic response to plant lectins and sensitivity to irradiation and various drugs (reviewed in Greaves *et al.*, 1973). Although it is difficult to distinguish resting T and B lymphocytes on morphological grounds by light or electron microscopy, a large number of surface differences have been demonstrated which have proved useful in distinguishing and separating the two classes of lymphocytes. Some of these surface differences can be recognized by antibody. For example, the θ alloantigen (defined by alloantibody made in one strain of mouse against thymocytes of another strain) is present on mouse thymocytes and T cells but not on B lymphocytes, and has proved to be a useful marker for T cells in mice (Raff, 1971). In addition, heteroantibodies made in rabbits against mouse or human T or B cells, or against known membrane determinants (e.g. Ig), after appropriate absorption, can be made to react specifically with one or other class of lymphocyte. Thus, anti-Ig antibody can be used to identify B cells in most, if not all species (Raff, 1971). Using antisera which react specifically with the surface of one or other lymphocyte type, one can kill off either cell population in the presence of complement, or alternatively, one can use antibody on digestible solid-phase immunoabsorbents (Schlossman and Hudson, 1973), or fluoresceinated antibody and fluorescence-activated electronic cell sorting (Hulett *et al.*, 1969) to purify either type of cell. Besides surface antigenic differences between T and B cells, the latter can bind antibody-antigen-complement complexes by means of surface complement receptors (Bianco *et al.*, 1970), and antibody-antigen complexes by means of receptors for the Fc part of complexed or aggregated Ig (Basten *et al.*, 1972). Since resting T cells do not express these receptors, they can be used to identify and separate B cells; there is evidence, however, that activated T cells may also have Fc receptors (Yoshida and Andersson, 1972). More recently, it has been shown that the majority of human thymocytes and T lymphocytes have surface receptors for sheep erythrocytes, while B cells do not, and this is now being widely used as a T cell marker in man (Jondal *et al.*, 1972). The usefulness of these various distinguishing surface antigens and receptors in studying the properties, functions and differentiation of T and B lymphocytes, should encourage the use of similar techniques in other biological disciplines, such as developmental biology.

ORIGINS AND DIFFERENTIATION OF LYMPHOCYTES

Immunocompetent lymphocytes develop in at least two stages, following which their further differentiation requires specific antigen. Firstly, hemopoietic stem cells differentiate into lymphocytes in the central lymphoid tissues (i.e. thymus, bursa of Fabricius, or "bursa equivalent"), and secondly, after further maturation, some of these lymphocytes migrate to the peripheral lymphoid tissues. In the peripheral tissues, immunocompetent T and B lymphocytes await appropriate encounter with their specific antigens which will induce them to differentiate into effector and/or "memory cells" (see below). A unique feature of lymphocyte development is the production of very large numbers of different clones, each capable of responding to a restricted range of antigenic determinants. This presumably explains the number of unique genetic mechanisms which have been shown to operate in their development (see below).

Hemopoietic stem cells

Hemopoietic stem cells are cells with a capacity for extensive proliferation and self-renewal which give rise to the specialized cells of the blood, including erythrocytes, granulocytes, monocytes, platelets and lymphocytes (Metcalf and Moore, 1971). They arise initially in the embryonic yolk sac, probably from cells that migrate out from the primitive streak (Murray, 1932), and subsequently seed the tissues that become sequentially the main sites of hemopoiesis, namely liver in the embryo and bone marrow in the adult. There is increasing evidence that both myeloid and lymphoid cells may be derived from a common stem cell (Wu *et al.* 1968) which in the appropriate microenvironmental milieu, gives rise to unipotential 'line-progenitor' cells, which still have extensive proliferative capabilities but are committed to differentiate into either myeloid or lymphoid cells (Metcalf and Moore, 1971). The question that remains to be answered is what is the level of commitment of the stem cells that enter the thymus and bursa (or bursa equivalent)? It seems likely that they are for the most part multipotential stem cells, but it has not been excluded that they are line-progenitor cells committed to becoming lymphoid or perhaps even committed to becoming T or B cells.

Lymphopoiesis in the central lymphoid organs

In most animals, lymphocytes first develop in the fetal thymus. The thymus anlage is composed of epithelial cells and derives from the third and fourth pharyngeal pouches. In mice (gestation = 20 days) the first stem cells, which appear to be large, basophilic blast-like cells, enter the thymus from the circulation around day 11 and the first small lymphocytes are seen by day 15 or 16 of embryonic life (Owen, 1972). This differentiation of large stem cells to small thymus lymphocytes is accompanied by the acquisition of a number of lymphocyte surface differentiation antigens (e.g. θ, Thymus-Leukemia /TL/ antigens) and can be followed *in vitro* (Owen and Raff, 1970). Thus, when 14 day fetal thymus, which contains no θ^+ or TL^+ lymphocytes, is cultured for 4 days, most of the cells are now lymphocytes and express θ and TL on their surface. In adult animals stem cells from bone marrow migrate to the thymus and differentiate to lymphocytes, but the magnitude of the traffic is much reduced compared to the fetus and newborn. In both fetus and adult, the differentiation of stem cells to thymus lymphocytes is probably induced by factors secreted by thymus epithelial cells, and recently it has been reported that a soluble factor(s) extracted from non-lymphoid elements of mouse thymus could induce θ^- and TL^- cells in adult bone marrow and spleen to express θ and TL alloantigens on their surface (Komura and Boyse, 1973), although it has not been shown that the responding cells are hemopoietic stem cells.

The initial development of B-type lymphocytes has only been studied in detail in birds, which are unique in having a discrete central organ of B-lymphocyte production, the bursa of Fabricius. The bursa arises as a sac-like evagination from the dorsal wall of the cloaca on day 5. Chromosome marker studies have shown that stem cells (morphologically identical to those seen in the chick fetal thymus) migrate from yolk sac to bursa beginning around day 12-13 and there differentiate to lymphocytes within 1-2 days (Owen, 1972). By day 14, bursa lymphocytes with IgM on their surface (presumably receptors) can be seen, while IgG-bearing bursa lymphocytes are seen a few days later (Kincade and Cooper, 1971). The nature of the inductive influence of the bursal microenvironment is unknown. In mammals, it is still unclear where B-type lymphocytes are produced, although in rodents at least (Everett and Caffrey, 1966) there is evidence for active lymphopoiesis within the hemopoietic tissues themselves (mainly liver in embryos and bone marrow in adults) and it seems likely that these tissues not only supply the stem cells for both T and B cell populations, but also are the sites where stem cells differentiate to B-type lymphocytes.

Further maturation and peripheralization of thymus and bursal lymphocytes

Using radioactive, chromosome and surface antigenic markers it has been established that lymphocytes migrate from thymus to peripheral lymphoid tissues to make up the T lymphocyte population (Owen, 1972). In mice, this process begins just before birth, with most of the seeding occurring in the first week of life, and being much reduced in adults. Thus neonatal thymectomy results in a marked T-cell deficiency, while thymectomy done later in life has much less effect. Such experiments indicate that the peripheral T cell population is largely self-sustaining and that the major role of the thymus is to produce lymphocytes and seed them to the peripheral lymphoid tissues. Nonetheless, there is evidence that the thymus can also influence a subpopulation of peripheral T cells by secreting humoral factors (Stutman *et al.*, 1969). Although it is clear that T cells derive from thymus lymphocytes, these two types of lymphocytes are very different in their properties. Most importantly, the great majority of thymus lymphocytes are immunologically incompetent (i.e. they cannot respond to antigen) while T cells are immunocompetent. Thus there must be another differentiation step between thymus lymphocytes and T cells. The finding of a small subpopulation (2-5%) of thymus cells, located in the thymus medulla, which is immunocompetent and has most of the properties of peripheral T cells (Owen, 1972) suggests that the second differentiation step may occur within the thymus and that T cell development may be visualized as stem cell ⟶ thymus lymphocyte ⟶ 'mature' thymus lymphocyte ⟶ peripheral T cell. However, it is becoming clear that such a scheme is almost certainly an oversimplification, for there is increasing evidence that cells may leave the thymus at different stages of maturation, or perhaps as distinct cell lines, giving rise to distinct subpopulations of peripheral T cells having different properties and functions (Raff and Cantor, 1971). That such subpopulations of T cells exist is no longer in doubt, but their functions and relationship to each other are still unknown.

Analogous to the situation in the thymus, isotopic labelling experiments have demonstrated that bursal lymphocytes migrate to peripheral lymphoid tissues, while observations on the effect of embryonic and adult bursectomy have suggested that the bursa's main function is to seed cells to the peripheral lymphoid tissues, mainly prior to and just after hatching, and that once seeded the bursa-derived B lymphocyte population is largely self-sustaining (Cooper *et al.*, 1972). The fact that, in ontogeny, bursal cells bearing IgM on their surface appear several days before surface IgG-bearing cells are seen, taken together with the finding that some bursal cells can be shown to produce both IgM and IgG (Kincade and Cooper, 1971), have suggested that all bursal lymphocytes pass through an IgM-bearing stage, with some cells later switching to IgG. This notion is supported by the recent findings that injecting anti-μ antibody (i.e. specific for the heavy chains of IgM) prior to hatching, combined with neonatal bursectomy, suppresses later production of IgG as well as IgM (Kincade *et al.*, 1970). Whether the IgM $\rightarrow$ IgG switch within an individual bursal (or bursal-derived) cell occurs only in the bursa and independently of antigen as concluded from experiments in chickens (Cooper *et al.*, 1972) or can be driven by antigenic stimulation, as suggested by experiments in mice (Pierce *et al.*, 1972) remains to be established.

The sequential maturation of B-type lymphocytes has been difficult to follow in mammals, where the bursa equivalent has not been definitively identified. However, it seems likely that many of the rapidly dividing lymphocytes in the hematopoietic tissues are immature bursa-equivalent lymphocytes, some of which further mature and acquire surface Ig receptors before migrating to the periphery.

Antigen-induced differentiation of peripheral lymphocytes

When antigen combines with its specific receptors on the surface of a T or B lymphocyte one of several things can happen to the cell: (1) it may be induced to divide and differentiate to an effector cell in some type of immune response (e.g. B cells to make antibody, T cells to secrete lymphokines, become "killer cells" or "helper cells"); (2) it may become *immunologically tolerant*, so that it will not be able to respond to a subsequent challenge with the same antigen; it is not known if such tolerant cells are eliminated or remain around in an inactive state; (3) it may be unaffected by the encounter. In addition, if the animal does respond to the antigen, on subsequent exposure to the same antigen, it will usually give a faster, greater and sometimes qualitatively different response, a phenomenon known as *immunological memory*. It is likely that memory involves both clonal expansion (i.e. division of virgin lymphocytes to give an increased number of cells able to respond on re-exposure to the antigen) and differentiation of virgin cells to memory cells, but it is not known whether memory cells are simply retired effector cells, cells at an earlier stage of differentiation than effector cells, or are derived by differentiation along a distinct memory pathway. All of these responses, including immunological tolerance and memory, are antigen specific and can involve T and/or B lymphocytes (Greaves *et al.*, 1973).

The "decision" that an individual lymphocyte makes on encounter with antigen, whether to "turn-on", "turn-off" or ignore, depends largely on the physical-chemical nature of the antigen and upon complex interactions with other lymphocytes and with macrophages. The way in which the interaction of antigen with receptors "signals" a lymphocyte is unknown. The recent discovery that lymphocyte receptors are able to move laterally in the plane of the membrane and can be induced to passively redistribute into clusters or actively gather over one pole of the cell forming a cap, if cross-linked by multivalent antigens (Taylor *et al.*, 1971; Loor *et al.*, 1972), suggests the possibility that signalling may involve receptor redistribution. Although there is no direct evidence for such an hypothesis, it would nicely explain the apparent requirement for multivalent binding when antigens, phytomitogens or antibodies directed against cell surface antigens induce lymphocyte proliferation and differentiation *in vitro* (Raff and de Petris, 1973).

It is of interest that lymphocytes can be activated to divide and differentiate into blast cells (which often have effector function) by a variety of nonspecific interactions at the cell surface, including the binding of a variety of plant lectins, bacterial lipopolysaccharides, antibodies against surface antigens and periodate (Greaves and Janossy, 1972). Frequently these substances only activate T cells (e.g. Concanavalin A, PHA) or B cells (lipopolysaccharide), even though they bind equally well to both types of cells (Greaves and Janossy, 1972), suggesting that the requirements for activation may be different for the two types of lymphocytes. In those cases where both T and B cells are activated by the same lectin (e.g. Pokeweed mitogen) the pattern of response is characteristic of the cell and not of the lectin (Greaves and Janossy, 1972). Thus B cells are induced to produce Ig and T cells to make lymphokines. These observations that lymphocyte activation can be induced by a variety of non-physiological cell surface interactions and that immunocompetent T and B lymphocytes are programmed to give distinctive but stereotyped responses regardless of the nature of the activating stimulus, have their counterparts in other biological systems. For example, in some forms of hyperthyroidism in man, autoantibody against thyroid follicular cells (called long-acting thyroid stimulator or LATS) induces these cells to produce excessive amounts of thyroglobulin, the response being indistinguishable from that induced by excessive amounts of the physiological stimulus, thyroid stimulating hormone (TSH) (Munro, 1971). Recently, it has been shown that Concanavalin A and wheat germ agglutinin are as effective as insulin in enhancing glucose transport in fat cells (Cuatracasas and Till, 1973). Thus it is likely that most mature, differentiated cells can be activated by a variety of non-sepcific, unphysiological stimuli under appropriate conditions. It would be of interest to see if relatively undifferentiated cells (e.g. stem cells) can be similarly induced to differentiate by non-physiological stimuli. Surface-binding lectins and antibodies may well prove to be useful probes for studying some aspects of differentiation in non-lymphoid systems.

Unique aspects of lymphocyte development

The characteristic feature of the immune system that distinguishes it from other physiological systems, is that it can respond specifically to a very large number of different antigens. As previously mentioned, the immune system manages this by producing many different lymphocyte clones, each able to respond to a restricted number of antigens. Although both T and B cells have been shown to be clonally restricted, only in the case of B cells have the receptors and secreted products been unequivocally characterized (and shown to be Ig), and I will confine my further discussion to B lymphocytes. There is recent evidence (Raff et al., 1973) that all of the Ig receptors on an individual B lymphocyte have the same specificity. /Although this has not been demonstrated for T cells, it seems a reasonable guess that they also express receptors of a single specificity. / This suggests that some time in their development, probably in the bursa or bursa-equivalent, and before encountering antigen, individual B-type lymphocytes become committed to making antibody of a single specificity, initially displayed on the surface as receptor and ultimately produced in large amounts and secreted if the cell is appropriately stimulated. To understand the molecular biological implications of this observation, and why unique genetic mechanisms operate in the development of the immune system, it is necessary to know something about Ig structure.

Ig molecules are constructed of two types of polypeptide chains, heavy (H) and light (L), arranged in symmetrical four-chain units, which occur either singly (IgG) or in multimers (IgA, IgM). Both L and H chains exist in various polymorphic forms, there being two types of L chain (λ and κ) and several classes of H chain (α, γ, μ, ϵ, δ, – corresponding to IgA, IgG, IgM, IgE, and IgD respectively), each divided into various subclasses. The N-terminal amino-acid sequences (105-120 residues) of either L or H chains are generally found to be unique for each different chain studied and are thus referred to as variable regions (V_L, V_H), while the remaining C-terminal residues are relatively constant and referred to as constant regions (C_L, C_H). Differences in the V regions (particularly in the hypervariable or "hot spot" regions) of both L and H chains account for differences in antibody specificity, as parts of V_L and V_H together make up the antigen-binding (i.e. combining) site. Thus a monomeric Ig molecule (LH LH) is bivalent with two identical combining sites. The class and subclass variation in C_H (see above) determines the biological activity of the secreted Ig in terms of its ability to fix complement, cross the placenta or adhere to the surface of various cell types, such as macrophages or mast cells.

Studies of the molecular biology of Ig synthesis have shown that, in general, Ig molecules are produced in the same way as other proteins, although there are a number of unique features. There is strong evidence that in each haploid genome a single pool of V_H genes is shared by all of the C_H genes, and that C_λ and C_κ genes have separate V_λ and V_κ gene pools (Pink et al., 1971), there being then three separate V gene pools. Thus, to form a single H or L chain, two separate genes (V_H and C_H, or V_L and C_L) are required; this being a unique example of two genes coding for one polypeptide chain (Hood and Talmage, 1970). There are reasons for believing that the VC link-up occurs at the DNA, rather than at the RNA or protein level, although the exact mechanism is unknown (Williamson, 1971). Teleologically it is not difficult to see the advantages of this genetic arrangement. Since both V_L and V_H contribute to combining site specificity, having separate V_L and V_H gene pools greatly increases the number of antibodies an animal can make. On the other hand, having the various classes and subclasses of C_H linked to the same V_H gene pool, allows individual B cell clones to produce antibodies with identical specificity yet having different biological properties, which a priori would seem advantageous.

It has long been known that a single antibody-secreting cell generally produces a uniform product, having one class, one light chain type and one specificity (Mäkelä and Cross, 1970). In addition, as discussed above, there is recent evidence that B cells, which are the precursors of antibody-forming cells, also express Ig receptors of a single specificity. Since lymphocytes are diploid and have both maternal and paternal V gene pools, it is clear that some special mechanism must operate to ensure that a B cell is monospecific. For if both maternal and paternal V genes were expressed in an individual B lymphocyte it is almost certain that a cell would make antibodies with at least two different specificities, the chances of the same V_L or V_H genes being transcribed from maternal and paternal chromosomes being remote. In fact, almost ten years ago it was observed that individual antibody-secreting cells from animals heterozygous for an Ig allotype (i.e. an Ig alloantigen) produce Ig of only one parental allotype, the gene coding for the other allotype apparently being inactive in that cell (Cebra *et al.*, 1966; Pernis *et al.*, 1965). Although both allotypes are found in the serum they are produced by different cells. This phenomenon is referred to as *allelic exclusion*, and has recently also been shown to apply to Ig receptors on B lymphocytes (Pernis *et al.*, 1970). It is not known if the entire chromosome or only the Ig genes are inactive. Allelic exclusion is thus far unique to immunoglobulin synthesis, although it is clearly similar to the Lyon phenomenon.

Thus, it appears that sometime in their development individual B lymphocytes become committed to expressing one V_H and one V_L gene, while the C_H gene expressed may vary during the lifetime of the clone. Since C_λ and C_κ have separate V gene pools, to maintain monospecificity, individual B cells must make either λ or κ L chains, but not both, and this seems to be the case. Finding B cells committed to making Ig of a single specificity is comforting, as the problem of determining how antigen activates a lymphocyte would be far more complicated if specificity selection at the level of an individual cell was also involved.

It is clear that in an immunologically mature animal that total V gene pool is very large, and perhaps the most fundamental question that remains to be answered in immunology is how this large gene pool, which determines antibody diversity, is generated. The two extreme views are represented by the *germ line theories* which maintain that all available V genes are present in each cell and are inherited, and the *somatic theories* which propose that an individual inherits few V genes and diversity is generated through some somatic process (e.g. mutation, recombination). In the latter case, each clone would have a different V gene pool. These and intermediate hypotheses have their proponents and the question remains open.

REFERENCES

1. Basten, A., Miller, J. F. A. P., Sprent, J. and Pye, J. (1972a). A receptor for antibody on B lymphocytes. I. Method of detection and functional significance. *J. Exp. Med. 135*:610-625.

2. Bianco, C., Patrick, R. and Nussenzweig, V. (1970). A population of lymphocytes bearing a membrane receptor of antigen-antibody complement complexes. I. Separation and characterization. *J. Exp. Med. 132*:702-720.

3. Brunner, K. T. and Cerottini, J.-C. (1971). Cytotoxic lymphocytes as effector cells of cell-mediated immunity. *In* Progress in Immunology (B. Amos, editor) 385-398. Academic Press, New York.

4. Burnet, F. M. (1959). The Clonal Selection Theory of Acquired Immunity. Cambridge University Press, London.

5. Cebra, J. J., Coberg, J. E. and Dray, S. S. (1966). Rabbit lymphoid cells differentiated with respect to λ, *a* and μ heavy polypeptide chains and to allotypic markers, Aa1 and Aa2. *J. Exp. Med. 123*:547-558.

6. Cooper, M. D., Lawton, A. R. and Kincade, P. W. (1972). A developmental approach to the biological basis of antibody diversity. *In* Contemporary Topics in Immunobiology (M. G. Hanna, editor), *1*:33-68. Plenum Press, New York.

7. Cozenza, H. and Köhler, H. (1972). Suppression of the antibody response by antibodies to receptors. *Proc. Natl. Acad. Sci. (U.S.A.) 69*:2701-2705.

8. Crone, M., Koch, C. and Simonsen, M. (1972). The elusive T cell receptor. *Transplant. Rev. 10*:36-56.

9. Cuatracasas, P. and Till, G. P. (1973). Insulin-like activity of Concanavalin A and wheat-germ agglutinin. Direct interaction with insulin receptors. *Proc. Natl. Acad. Sci. (U.S.A.) 70*:485-489.

10. Everett, N. B. and Caffrey, R. W. (1966). Radioautographic studies of bone marrow small lymphocytes. *In* The Lymphocyte in Immunology and Hemopoiesis (J. M. Yoffey, editor) 108-119. Arnold, London.

11. Greaves, M. F. and Janossy, G. (1972). Elicitation of selective T and B lymphocyte responses by cell surface binding ligands. *Transplant. Rev. 11*:87-130.

12. Greaves, M. F., Owen, J. J. T. and Raff, M. C. (1973). T and B lymphocytes: origins, properties and roles in immune responses. *Excerpta Medica* Amsterdam – in press.

13. Hood, L. and Talmage, D. W. (1970). Mechanism of antibody diversity: germ line basis for variability. *Science 168*:325-333.

14. Hulett, H. R., Bonner, W. A., Barret, J. and Herzenberg, L. A. (1969). Cell sorting: automated separation of mammalian cells as a function of intracellular fluorescence. *Science 166*:747-748.

15. Jondal, M., Holm, G. and Wigzell, H. (1972). Surface markers on human T and B lymphocytes. I. A large population of lymphocytes forming non-immune rosettes with SRBC. *J. Exp. Med. 136*:207-215.

16. Kincade, P. W. and Cooper, M. D. (1971). Development and distribution of immunoglobulin-containing cells in the chicken. *J. Immunol. 106*:371-382.

17. Kincade, P. W., Lawton, A. R., Bockman, D. E. and Cooper, M. D. (1970). Suppression of immunoglobulin G synthesis as a result of antibody-mediated suppression of immunoglobulin M synthesis in chickens. *Proc. Natl. Acad. Sci. (U.S.A.) 67*:1918-1925.

18. Kumuro, K. and Boyse, E. A. (1973). In vitro demonstration of thymic hormone in the mouse by conversion of precursor cells into lymphocytes. *Lancet* i, 740-742.

19. Lawrence, H. S. and Landy, M. (Editors) (1969). Mediators of Cellular Immunity. Academic Press, New York.

20. Loor, F., Forni, L. and Pernis, B. (1972). The dynamic state of the lymphocyte membrane. Factors affecting the distribution and turnover of surface immunoglobulin. *Eur. J. Immunol. 2*:203-212.

21. Mäkelä, O. and Cross, A. M. (1970). The diversity and specialization of monocytes. *Progr. Allergy 14*:145-207.

22. Metcalf, D. and Moore, M. A. S. (1971). Hemopoietic cells: their origin, migration and differentiation. Frontiers of Biology, Vol. 24. North Holland, Amsterdam.

23. Mitchison, N. A., Rajewsky, K. and Taylor, R. B. (1970). Co-operation of antigenic determinants and of cells in the induction of antibodies. *In* Developmental

Aspects of Antibody Formation and Structure. (J. Sterzl and H. Riha, editors) 547-561. Academic Press, New York.

24. Munro, D. S. (1971). The long-acting thyroid stimulator. *In* Current Topics in Experimental Endocrinology, 1: 175-198.

25. Murray, P. D. F. (1932). The development in vitro of the blood of the early chick embryo. *Proc. Roy. Soc. B, 111*:497-521.

26. Owen, J. J. T. (1972). The origins and development of lymphocyte populations. *In* Ontogeny of Acquired Immunity (A Ciba Foundation Symposium), 35-54. Elsevier-Excerpta Medica - North Holland, Amsterdam.

27. Owen, J. J. T. and Raff, M. C. (1970). Studies on the differentiation of thymus-derived lymphocytes. *J. Exp. Med. 132*:1216-1232.

28. Pernis, B., Chiappino, G., Kelus, A. S. and Gell, P. G. H. (1965). Cellular localization of immunoglobulin with different allotypic specificities in rabbit lymphoid tissues. *J. Exp. Med. 122*:853-876.

29. Pierce, C. W., Solliday, S. M. and Asofsky, R. (1972). Immune response in vitro. IV. Suppression of primary γM, γG and γA plaque-forming cell responses in mouse spleen cell cultures by class-specific antibody to mouse immunoglobulins. *J. Exp. Med. 135*:698-717.

30. Pernis, B., Forni, L. and Amante, L. (1970). Immunoglobulin spots on the surface of rabbit lymphocytes. *J. Exp. Med. 132*:1001-1018.

31. Pink, R., Wang, A. C. and Fudenberg, H. H. (1971). Antibody variability. *Ann. Rev. Med. 22*:145-170.

32. Raff, M. C. (1971). Surface antigenic markers for distinguishing T and B lymphocytes in mice. *Transplant. Rev. 6*:52-80.

33. Raff, M. C. and Cantor, H. (1971). Subpopulations of thymus cells and thymus-derived lymphocytes. *In* Progress in Immunology (B. Amos, editor) 83-93. Academic Press, New York.

34. Raff, M. C. and de Petris, S. (1973). Movement of lymphocyte surface antigens and receptors: the fluid nature of the lymphocyte plasma membrane and its immunological significance. *Fed. Proc. 32*:48-54.

35. Raff, M. C., Feldmann, M. and de Petris, S. (1973). Monospecificity of bone marrow-derived lymphocytes. *J. Exp. Med. 137*:1024-1030.

36. Shevach, E. M., Paul, W. E. and Green, I. (1972). Histocompatibility-linked immune response gene function in guinea pigs: specific inhibition of antigen-induced lymphocyte proliferation by alloantisera. *J. Exp. Med. 136*:1207-1221.

37. Schlossman, S. F. and Hudson, L. (1973). Specific purification of lymphocyte populations on a digestible immunoabsorbent. *J. Immunol. 110*:313-315.

38. Singhal, S. K. and Wigzell, H. (1971). Cognition and recognition of antigen by cell-associated receptors. *Progr. Allergy 15*:188-222.

39. Stutman, O., Yunis, E. J. and Good, R. A. (1969). Carcinogen-induced tumor of the thymus. IV. Humoral influence of normal thymus and functional thymomas and influence of post-thymectomy period of restoration. *J. Exp. Med. 130*:809-819.

40. Taylor, R. B., Duffus, W. P. H., Raff, M. C. and de Petris, S. (1971). Redistribution and pinocytosis of lymphocyte surface immunoglobulin molecules induced by anti-immunoglobulin antibody. *Nature New Biol. 233*:225-229.

41. Williamson, A. R. (1971). Biosynthesis of antibodies. *Nature, Lond. 231*:359-362.

42. Wu, A. M., Till, J. E., Siminovitch, L. and McCullogh, E. A. (1968). Cytological evidence for a relationship between hematopoietic colony-forming cells and cells of the lymphoid system. *J. Exp. Med. 127*:455-464.

43. Yoshida, T. O. and Andersson, B. (1972). Evidence for a receptor recognizing antigen complexed immunoglobulin on the surface of activated mouse thymus lymphocytes. *Scand. J. Immunol. 1*:401-408.

EMBRYONIC AND FETAL ANTIGENS IN CANCER CELLS[1]

Joseph H. Coggin, Jr.
and
Norman G. Anderson

The University of Tennessee, Department of Microbiology
Knoxville, Tennessee
and
The Molecular Anatomy (MAN) Program, Oak Ridge National Laboratory[2]
Oak Ridge, Tennessee

With few exceptions cancer cells are recognized to have novel antigens by the immune system of the host (Old, *et al.*, 1964; Klein, 1966). Several dozen reports indicate that the humoral and the cell mediated arms of the immune response are activated against these neoantigens in the cancers of men and the mouse. This is generally the case whether one considers spontaneously occurring, virally or chemically-induced neoplasms of the rodents which have been tested to date. These observations suggest two fundamental questions about the character of antigens on cancer cells. Where do the "new" antigens come from (intrinsic or extrinsic DNA) and, when present, why do they seem ineffectual in arousing the host to successfully eliminate the neoplasm which exhibits them? In the course of this presentation we propose that the explanations to both questions may be closely related and that the biological role of these antigens in normal fetal development and in cancer could be similar.

Until recently cancer cells of laboratory animals were believed to contain only antigenic entities of a unique or tumor specific character. Chemically-induced tumors of rodents, for example, were thought to possess individually specific, non-cross reactive membrane antigens (Klein, *et al.*, 1960; Old, *et al.*, 1968) to which the host responded. Similarly, tumors induced by a given oncodnavirus possessed antigens uniquely present on cancer cells transformed

[1]Supported by The Virus Cancer Program, National Cancer Institute (CP-73-210) and The Atomic Energy Commission (AT 40-1 3646).
[2]Operated by Union Carbide Corporation for the U. S. Atomic Energy Commission.

by the specific virus and not present on tumors induced by other oncodna-viruses or on normal tissues (see Haughton and Nash, 1969). It was unclear whether the surface antigens present on the membranes of a variety of cancers which arose spontaneously in rodents cross-reacted. What was clear was that immune reactivity to these antigens was of the antibody *and* the cell-mediated, transplantation type.

In contrast, human cancers now are generally recognized to have cross-reactive antigens shared between tumors of common histologic origin (Morton, 1970; Hellstrom *et. al.*, 1971). Individually specific antigens, termed tumor specific transplantation antigens or TSTA's may exist on human tumors but the outbred character of human subjects prohibits this immunologic discrimi-nation at present. Cross-reactive antigens can be detected despite this diffi-culty. It is of particular importance to understand the source of these cross-reacting, histologically-related tumor antigens appearing on carcinomas, sarcomas and blood-borne cancers of humans. Are such antigens derived from mutational changes in the genetic information of the initial cancer cell or from exogenous DNA or RNA present in oncogenic viruses? Is it possible that the genetic potential for producing these cross-reacting tumor neoantigens innately resides in the normal adult cell genome in an untranscribed, silent form? This possibility has been of central concern to us in our present research. If tumor cells indeed possess fetal antigens we must then give major attention to the relationship between the expression of fetal macromolecules and the cancerous traits of the tumor cells which display these antigens.

Model System Studies. It seems fruitful at this juncture to briefly review a virus cancer model system employed for a number of years now in our laboratories in an effort to collect data which may afford insight into these important questions in cancer biology. We have worked with the oncodna-viruses of the adenovirus group (serotypes 7 and 31) and with the papova-virus, simian virus 40 (SV40). These viruses were previously described to be tumor-producing or oncogenic when injected subcutaneously into the sub-scapular space of newborn hamsters (Goldner, *et al.*, 1964; Coggin *et al.*, 1967). Cancers (fibrosarcomas) induced by these viruses appear at the site of virus infection with high frequency many weeks to months later and most tumors are observed to be virus free at the time of detection (Table 1). The cancers are generally malignant invading many vital organs, especially the lungs.

Immunization against tumor induction in this system could be achieved by the subsequent administration of additional "live" virus after the hamsters achieved immunological competence (3-5 weeks of life) or by injection of radiation-inactivated homologous tumor tissue or tumor cells propagated *in vitro* (Goldner, *et al.*, 1964; Coggin, *et al.*, 1967). The protective effect was specific. SV40 would only interfere with SV40 oncogenicity. Adenovirus 31

TABLE 1.

The Induction of Fibrosarcomas in Inbred Syrian Golden Hamsters (Strain LSH) Following Neonatal with SV40 or Adenovirus 31

Week Post-Infection with Virus	Percentage of Hamsters Developing Tumors			
	SV40 Infected		Adenovirus 31 Infected	
	Male	Female	Male	Female
0	0	0	0	0
3	0	0	2	5
9	0	0	29	35
15	5	2	60	55
20	30	30	95	95
30	67	75	100	95
40	90	85	100	100
50	100	100	100	100

or tumor produced by it would only cause the rejection of adenovirus 31 tumors but not SV40 or other adenovirus induced tumors (Coggin, *et al.*, 1969; Ambrose, *et al.*, 1969). Hence, the viral transformed cells were said to possess tumor specific transplantation antigens (TSTA's) specifically induced by the homologous, transforming agent.

For several years we tried to isolate soluble TSTA's free from these tumor cell types. Early efforts were negative and generally yielded materials, released from the tumors, which caused enhancement or more rapid tumor appearance (Coggin, *et al.*, 1967; Coggin, *et al.*, 1967a). Membranes from both SV40 or adenovirus tumors, however, were capable of inducing the desired protection against challenge with homologous live tumor cells in vaccinated animals (Coggin, *et al.*, 1969) with some reliability.

What is significant from the above information for this consideration is that each tumor had a virus-related TSTA capable of conferring tumor immunity against other tumors produced by that virus but ineffective against tumors produced by other unrelated viruses. Comparable antigens were not observed in normal adult hamster tissues or organs nor in the normal tissues and organs of other adult rodents. Significantly, mouse cells transformed *in vitro* by SV40 (mice do not develop SV40 neoplasms when infected at birth with SV40) possess TSTA's which will protect hamsters against SV40 onco-genicity suggesting that the transforming virus codes for or induces identical TSTA's in cells of different rodent species and in human cells as well (Girardi, 1965; Kit, *et al.*, 1969). Hence, human, mouse and hamster cells transformed by SV40 possess identical TSTA.

Fetal Antigens in SV40 Tumors. In 1969 we reported the transient occurrence of an immunoglobulin in the serum of pre-tumor bearing animals or in the serum of animals rendered hyperimmune to SV40 neoplasms by immunization with SV40 virus or tumor cells which reacted specifically with membrane antigens on SV40 tumor cells (Coggin *et al.*, 1970; Ambrose *et al.*, 1971a). The functional interaction of this antibody with the tumor cells was cytostatic and rarely cytotoxic. We initially believed that the immunoglobulin was directed against TSTA-related determinants present on the tumor cell because of the parallel specificity observed between the occurrence of tumor immunity and the presence of circulating C antibody. Injection of normal hamster tissue into syngeneic hamsters did not elicit a similar immunoglobulin nor did heterologous tumors. The later observation that a similar antibody occurred transiently during normal pregnancy in hamsters (Coggin, *et al.*, 1970), disappearing shortly after birth (Table 2), was most disturbing. A similar observation by others stimulated an examination of whether antigenic determinants on embryonic tissues might serve to elicit transplantation-like reaction against SV40 tumor cell or SV40-induced, autochthonous neoplasms *in vivo* (Duff and Rapp, 1970). The results were encouraging. Fetal cells of the mouse, hamster or human in early to late gestation (Coggin, *et al.*, 1970; Ambrose, *et al.*, 1971: Coggin, *et al.*, 1971) were effective inducers of significant, albeit modest, protection against SV40 and adenovirus tumors in syngeneic hamsters. These and other findings (Brawn, 1970; Baldwin, *et al.*, 1972; Herberman, *et al.*, 1971, LeMevel and Wells, 1973) demonstrate that many tumors of rodents possess fetal antigens reexpressed in oncodnavirus,

TABLE 2.

Detection of C Antibody in Multiparous Female Hamsters Following Conception

Days Post Conception[a]	Cytostasis Against SV40 Tumor Cells[b]
Virgins[c]	−
9	−
14	+
18	+
19	+
20	−

[a]Chambers containing 15,000 viable SV40 tumor cells were placed into the peritoneal cavity 5 days before the indicated days below.

[b]Percentage cytostasis was determined by removing the chamber at 5 days implantation at the indicated dates from 10 hamsters and determining the average number of viable cells in the pregnant hamsters or mothers. This figure was compared with cell counts from chambers implanted simultaneously into virgin hamsters and counted on the same sampling date to determine the percentage of cytostasis.

[c]Virgin hamsters were the same age as the multiparous animals and cell counts over 5 days in these animals are identical to those obtained in male hosts (550,000 viable cells).

chemically-induced, and spontaneous tumors (Table 3). Some of the more intriguing facts we have learned about fetal antigen expression on tumor cells and their antigenic characteristics are summarized in Table 4. These data indicated that tumor cells of many types possessed cross-reacting fetal antigens and, perhaps in addition, TSTA's as well. We currently do not know how many different fetal antigens there are, nor do we understand the mechanism for triggering their appearance in many cancer cells.

A number of investigators had previously noted cross-reactions between fetal tissues and tumors and some of this work is summarized in Table 5. Several investigators have attempted immunization against tumors in other rodents (rats and mice) with poor success (Buttle, *et al.*, 1964; Blair, 1970; Ting, 1968; Ting, *et al.*, 1971), but in each case one or more of the guidelines for achieving activated cell-mediated immunity had not been observed. Recent successes in the mouse system with chemically-induced tumors (Bendich, *et al.*, 1973; LeMevel and Wells, 1973) and in the guinea pig (Grant, *et al.*, 1973)

TABLE 3.

Tumors Known to Possess Embryonic or Fetal Surface Antigens
(Syngeneic or Autochthonous Models Only)

Animal Model: Tumor Type	Tumor Origin (No. of tumors)	Immune Reaction Detected: *In Vitro*	*In Vivo*
Hamster:			
Sarcoma	Spontaneous	+	+
	SV40 Virus	+	+
	Adenovirus 7	+	+
	Adenovirus 31	+	+
	Polyoma Virus	+	+
	Maloney Virus	+	+
Carcinoma	Methylcholanthrene (2)	+	+
	DMBA	+	+
	SV40	+	+
Lymphoma	SV40	+	?
Rat:			
Sarcoma	MCA - (10)	+	+
	DMBA - (4)	?	+
Mouse:			
Sarcoma	SV40 (MKS-A)	+	?
	MCA (3)	+	+
Leukemia	Rauscher	+	+
Lymphoma	?	+	+
Carcinoma	MCA	+	+
Plasma Cell	Spontaneous	?	+
Ascites	L-cell	?	+
Guinea Pig:			
Carcinoma	Hydrocarbon (8)	+	+

TABLE 4.

A Summary of Facts Relating to the Biological and Immunologic Traits of Fetal Antigens on Viral Transformed Hamster Tumor Cells.

Biological:

    1. Immunization of adult hamsters with syngeneic fetal cell suspensions can be achieved in male recipients.

    2. Adult female hamster recipients of syngeneic fetal vaccines respond only with antibody responses and do not demonstrate transplantation resistance to tumor following exhaustive immunization efforts.

    3. Fetal cells must be inactivated by irradiation prior to injection for maximal immunization.

    4. Suitable immunization can only be obtained with midgestation fetal cell homogenates; late gestation material is non-immunogenic.

    5. Fetus must come from primaparous donors, not from multiparous donors.

    6. Human and mouse fetus will serve to protect hamsters against SV40 oncogenicity.

    7. Pregnant females develop transient humoral *and* cellular immunity to tumor during the course of pregnancy and both immune responses disappear shortly after birth.

Immunologic:

    1. Antigen is not present after the 11th day of gestation in the hamster but is present at day 10.

    2. Whole fetal cells, freshly collected, irradiated and injected are essential for routine immunization against SV40 tumors.

    3. Soluble fetal antigen induces humoral antibody but does not elicit cell mediated immunity.

    4. Fetal antigens are autosoluble from fetal cells and from tumor cells.

    5. Fetal antigens are not Forssman or heterophile antigens and do not appear to be organ specific antigens.

    6. Serum from normal pregnant (multiparous) animals blocks tumor specific cytotoxicity by sensitized lymph node cells (LNC) and serum from tumor-bearing hamsters blocks the cytotoxicity of multiparous LNC for SV40 tumor target cells.

show that protection against tumor induction by previous immunization with irradiation fetal tissues may be a general phenomenon provided one adheres to certain "rules" in preparing the vaccine. To date we have not found any tumors of rodents which do not respond to some degree to fetal tissue sensitization, although we cannot yet demonstrate (have not exhaustively tested) a broad spectrum of cross-protection with different fetal *species* in rodents other than the hamster. The essential point here is that many model tumors possess fetal antigens and these antigens can and do elicit both humoral and cellular immune responses in the cancer-bearing host.

    These fetal antigens now exist presumably in addition to TSTA's on the tumor cell membranes. It is still possible that even the TSTA's represent aberrant expression of new phenotypes of fetal or embryo-associated macromolecules under specific viral regulation. Human tumors are generally recognized to possess cross-reacting antigens (by histologic type) but not all of

TABLE 5.
Summary of Efforts to Detect Embryonic or Fetal Antigens in Model Tumor Systems Prior to 1970.

Date	Model	Fetal Antigen Demonstrated	Comment	Reference	Assay	Allogeneic	Syngeneic or Inbred
1906	Mice-spontaneous sarcoma (?)	+	Fetal immunization prevented tumor transplantation	Schone, 1906	TR	+	
1962	Rat-human sarcoma H51	+	Cortisone treated rats	Buttle, et al., 1962	TR[1]	+	
1964	Mice-chemically-induced sarcomas	+	Only detected with non-syngeneic fetus	Buttle, et al., 1964, 1967	TR	+	
1967	Mice-3-MCA sarcomas	+	Moderate to weak protection	Prehn, 1967	TR		+
1968	Mice-polyoma tumor	+	Alloantisera used	Pearson, et al., 1968	AS[2]	+	
1968	Mice-polyoma tumor	−	No protection observed against polyoma tumor challenge	Ting, 1968	TR		+
1970	Mice-72 mouse tumors	+	Many tumors cross-react with anti-fetal sera	Stonehill, et al., 1970	AS	+	
1970	Pregnant mice-3-MCA sarcomas	+	Pregnant effector cells destroy tumor cells	Brawn, 1970	MC[3]		+
1970	Mice-MTV	-(?)	Attempted to prevent MTV tumors	Blair, 1970	TR		+
1970	SV40 - unfertilized mouse ova	+	Alloantisera reacted against SV40 tumor cells	Baranska, et al., 1970	AS	+	

1. Tumor rejection assay.
2. AS = Allogenic serum prepared against mouse fetus, absorbed and tested.
3. MC = Microcytotoxicity test.

these are known to be of fetal origin at present and it is most important that this determination be made. The cross-reacting S_2 antigens on human cancers of similar histologic origin elicit antibodies which could be absorbed away be fetal tissues (Mukherji, *et al.*, 1973).

Materno-fetal Relationships, Pregnancy and Tumors. To our thinking, the discovery by Brawn (Brawn, 1970) that inbred, pregnant mice possessed cytotoxic lymph node cells (LNC's) against several chemically-induced tumors was a most significant observation. In the natural immunologic "conversation" between mother and fetus, the mother becomes sufficiently sensitized to the anlagen of fetal or embryonic antigens (not conventional transplantation antigens in this situation since inbred animals were used) to show cross-reactive cytotoxicity against a variety of tumors *in vitro*. Humoral responses are engendered as well (recall Table 2). We have confirmed Brawn's observation that LNC's from pregnant rodents are cytotoxic for a spectrum of both virally-induced (Girardi, 1973) and chemically-induced tumors. Baldwin and his associates have made similar observations (Baldwin, *et al.*, 1973).

We do not know the range of immunologic sensitization to fetal antigens that occurs in pregnancy, nevertheless one is impressed that it must be great since all the tumors of mice, hamsters and more recently rats, selected totally by convenience (and thus chance), have shown uniform sensitivity to LNC's from pregnant donors. Since many of the tumors employed do not induce cross-protection against each other, yet the tumors cross-react with fetal antigens, we must conclude that several to many fetal antigens exist, perhaps related only by the histologic origin of the cancers. This observation would seem to suggest that an undetermined spectrum of fetal antigens exists. Supportive data (Ting, *et al.*, 1972) for this contention have been obtained in an SV40 mouse model system using isotopically-labeled antibody. We believe that these data are most interesting but must be regarded with caution since successful (lethal) tumors in nature generally possess much weaker antigens. Only highly antigenic neoplastic cells which are either non-oncogenic or weakly oncogenic in adult mice yield similar results using the antiglobulin test procedure of Ting.

In our laboratory we were recently able to transfer immune protection against SV40 tumor challenge to normal, syngeneic hamsters by subcutaneous injection of LNC's from pregnant (primaparous or multiparous) donors (see Girardi, *et al.*, 1973). This method seems to be the easiest way to test individual tumors for fetal antigens and avoid the many problems associated with obtaining and preparing x-irradiated fetal homogenates for direct immunization.

Temporal Expression of Fetal Antigens in Normal Development. It is conceivable that fetal antigens are present on germinal cells prior to conception. Baranska, *et al.*, (1970) have suggestive evidence that this is the situation in unfertilized mouse eggs. Her results were obtained with allogeneic antibody and the antigens on the eggs were not examined for their capacity to induce cell mediated immunity. Do unfertilized hamster eggs possess fetal antigens? Previous work had shown clearly that the fetal antigens cross-protective against SV40 tumors was silenced between the 10th and the 11th day of gestation (Coggin, *et al.*, 1970; Hannon, *et al.*, 1973). An approach was selected to test unfertilized eggs of syngeneic hamsters *in situ* in the ovaries for their capacity to induce protection against SV40 tumors. Young adult female hamsters were subjected to either true ovariectomy where the ovaries were surgically removed or to simple ligation of the ovary where a tight suture was placed around the base of the ovary and the ovary left in place. It was hoped that in the course of resorption of the ligated organ, autosensitization might occur to any fetal determinants on the eggs. Three immunologic test parameters were employed to test for sensitization of the cell mediated immune system to fetal autoantigens if they were present in the resorbed eggs. Peritoneal exudate cells (PEC) were collected from each of these two groups and were tested for their capacity to destroy SV40 hamster tumor target cells known to bear fetal antigens *in vitro* in the microcytotoxicity test. Similar PEC's were tested for their protective ability when admixed with SV40 tumor cells *in vitro* and subsequently administered subcutaneously to the normal male test hamsters. Lastly, donor animals were tested for their ability to resist a direct SV40 tumor cell challenge. The results demonstrated quite clearly by all three test indicators that indeed the ovary tissues containing the unfertilized eggs did possess fetal antigens cross-protective against SV40 tumors with similar fetal determinants. Challenged animals whose ovaries were ligated and were observed to resorb did reject tumors with reproducibility (Winslow, *S. G.*, personal communication) more than 50% of the time. Females whose ovaries were ligated removed surgically were as sensitive to SV40 tumor challenge as normal female test animals. Cytotoxic action of the PEC's obtained *in vitro* in the microcytotoxicity assay or by passive transfer *in vivo* paralleled these findings with only the females subjected to ligation yielding cytotoxic or protective effector cells (PEC's) against SV40 tumor cells. These data lead us to conclude that unfertilized hamster eggs did indeed bear fetal antigens. We are currently seeking to determine whether ligation of the ovary will prevent the occurrence of spontaneous neoplasms of mice and autochthonous tumors induced in neonatally infected hamsters with one of several oncodnaviruses.

Immunology of Fetal Development and the Cancer Analogy. In previous work we have demonstrated that factor(s) in the serum of SV40-tumor bearing hamsters and in the serum of pregnant hamsters (multiple pregnancy donors) can interfere with or block the destruction of SV40 target cells by effector LNC's or PEC's from either SV40-tumor immune or multiparous hamsters, respectively (see Anderson and Coggin, 1973). Recently, C. Young-Whitley (1973) in our laboratory demonstrated that serum from *pregnant* multiparous hamsters could block the cytotoxic destruction of SV40 target tumor cells by effector LNC's sensitized to SV40 tumor. This cross-blocking relationship is most significant in our view and indicates that fetal antigens may play a considerable role in tumor progression. If tumor cells possess fetal antigens which elicit blocking antibody (efferent and/or afferent interference) or if these antigens complexed with antibody in the plasma of tumor-bearing animals to block the destructive action of sensitized effector cells (efferent interference) which might have otherwise served to promote tumor rejection, then fetal antigens should be of major interest to oncologist. Quantitative monitoring of such antigens should be possible since similar entities (the carcinoembryonic antigens) can routinely be detected in human cancer patients (see Zamcheck, 1972) and such indices may be useful in determining the status of tumor growth. If fetal antigens from tumors are generally soluble into the fluids and tissues of the host they may interfere with tumor elimination by yet another mechanism similar to that described initially by us (see Coggin, *et al.*, 1973) and demonstrated quite elegantly by Currie and Basham (1972) in rats and humans with cancer. In those studies, substances (fetal antigens?) could be washed from the surfaces of non-cytotoxic effector cells of tumor-bearing hamsters, rats or humans rendering the cells selectively cytotoxic for tumor target cells but not for "normal" cells. In recent studies yet unpublished, we have similarly observed that effector cells from multiparous female rodents are not normally cytotoxic for adenovirus or SV40 tumor target cells unless thoroughly washed before introduction into the test. Since washing of the effector cells is a *routine* requirement in the preparation procedure to achieve uniform effector cell counts it is not surprising that many investigators have not recognized that tumor-bearer effector cells from donors with large tumors are not cytotoxic in the host but can be so rendered by the washing procedure performed as a part of the microcytotoxicity test.

The materials removed by washing may be fetal antigens in circulation in *excess* in the body fluids of the tumor-bearing donors. This experimental point must be investigated and, if true, must be documented for human cancer situations. Detection of these antigens could then be used as a detection tool in early cancer diagnosis and, as mentioned previously, could be especially valuable in monitoring the efficacy of cancer therapy.

In a slightly different vein, let us consider for a moment other implications from the growing recognition that many and perhaps most tumors exhibit fetal macromolecules reexpressed at their surfaces as well as intracellularly. Is reexpression of "fetal" genetic information (retrogenesis) essential to neoplastic transformation? Can one find enzyme complexes and their percursor RNA transciptional products which must be reactivated through induction or derepression by carcinogens before cancer can occur? Can we identify fetal or embryo-specific proteins which correlate with malignant cell transformation and is it possible that selective chemotherapeutic interference with the synthesis of these embryo-specific substances from intrinsic DNA can be selectively achieved to destroy cancers; empirical chemotherapeutic trial and error approaches have generally failed against most forms of cancer. It is possible that knowledge of the specific synthesis to be inhibited (e.g., embryonic specific) might yield a more promising approach. Is it possible to immunize individuals against spontaneous cancer using fetal antigens without detrimental, autoimmune side-effects? We can only speculate at this time but none of these approaches can be disregarded pending further acquisition of knowledge about the relationship between retrogression and cancer. What is exciting is that avenues for experimental exploitation are now obvious and tangible and this is most refreshing and encouraging.

REFERENCES

1. Ambrose, K. R., Candler, E. L., and Coggin, J. H. (1969). Characterization of tumor-specific transplantation immunity reactions in immunodiffusion chambers *in vivo*. *Proc. Soc. Exp. Biol. and Med. 123*, 1013-1020.

2. Ambrose, K. R., Anderson, N. G., and Coggin, J. H. (1971). Interruption of SV40 oncogenesis with human fetal antigen. *Nature 233*, 194-195.

3. Ambrose, K. R., Anderson, N. G., and Coggin, J. H. (1971a). Cytostatic antibody and SV40 tumor immunity in hamsters. *Nature 233*, 321-325.

4. Baldwin, R. W., Glaves, D., and Vose, B. M., (1972). Embryonic antigen expression in chemically induced rat hepatomas and sarcomas. *Int. J. Cancer 10*, 233-243.

5. Baranska, W., Koldovsky, P., and Koprowski, H. (1970). Antigenic study of unfertilized mouse eggs: Cross reactivity with SV40-induced antigens. *Proc. Nat. Acad. Sci. U.S.A. 67*, 193-199.

6. Bendich, A., Borenfreund, E., and Stonehill, E. H. (1973). Protection of adult mice against tumor challenge by immunization with irradiated adult skin or embryo cells. *J. of Immunol. 111*, 284-285.

7. Blair, P. B. (1970). Search for cross-reacting antigenicity between mammary tumor virus-induced mammary tumors and embryonic antigens: effect of immunization on development of spontaneous mammary tumors. *Cancer Res. 30*, 1199-1202.

8. Brawn, J. R. (1970). Possible association of embryonal antigen(s) with several primary 3-methylcholanthrene-induced murine sarcomas. *Int. J. Cancer 6*, 245-249.

9. Buttle, G. A. H., Eperon, J. L., and Kovacs, E. (1962). An antigen in malignant and in embryonic tissues. *Nature (London) 194*, 780.

10. Buttle, G. A. H., Eperon, J. and Menzies, D. N. (1964). Induced tumor resistance in rats. *Lancet 7349*, 12-14.

11. Buttle, G. A. H., and Frayn. (1967). Effect of previous injections of homologous embryonic tissue on the growth of certain transplantable mouse tumors. *Nature 215*, 1495-1497.

12. Coggin, J. H., Larson, V. M., and Hilleman, M. R. (1967). Prevention of SV40 virus tumorigenesis by irradiated, disrupted, and iododeoxyuridine treated tumor cell antigens. *Proc. Soc. Exp. Biol. and Med. 124*, 774-784.

13. Coggin, J. H., Larson, V. M., and Hilleman, M. R. (1967a). Immunologic responses in hamsters to homologous tumor antigens measured *in vivo* and *in vitro*. *Proc. Soc. Exp. Biol. and Med. 124*, 1295-1302.

14. Coggin, J. H., Elrod, L. H., Ambrose, K. R., and Anderson, N. G. (1969). Induction of tumor-specific transplantation immunity in hamsters with cell fractions from adenovirus and SV40 tumor cells. *Proc. Soc. Exp. Biol. and Med. 132*, 328-336.

15. Coggin, J. H., Ambrose, K. R., and Anderson, N. G. (1970). Fetal antigen capable of inducing transplantation immunity against SV40 hamster tumor cells. *J. Immunol. 105*, 524-526.

16. Coggin, J. H., Ambrose, K. R., Bellomy, B. B., and Anderson, N. G. (1971). Tumor immunity in hamsters immunized with fetal tissues. *J. Immunol. 107*, 526-533.

17. Coggin, J. H., Ambrose, K. R., and Anderson, N. G. (1973). Phase specific surface autoantigens on membranes of fetus and tumors. *Adv. Exptl. Med. and Biol. 29*, 483-490.

18. Currie, G. A. and Basham, C. (1972). Serum mediated inhibition of the immunological reactions of the patient to his own tumor: A possible role for circulating antigen. *Br. J. Cancer 26*, 427-438.

19. Duff, R., and Rapp, F. (1970). Reaction of serum from pregnant hamsters with surface of cells transformed by SV40. *J. Immunol. 105*, 521-523.

20. Girardi, A. J. (1965). Prevention of SV40 virus oncogenesis in hamsters, I. Tumor resistance induced by human cells transformed by SV40. *Proc. Nat. Acad. Sci. U.S.A. 54*, 445-451.

21. Girardi, A. J., Reppucci, P., Dierlam, P., Rutala, W., and Coggin, J. H. (1973). Prevention of Simian virus 40 tumors by hamster fetal tissues: Influence of parity status of donor females on immunogenicity of fetal tissues and on immune cell cytotoxicity. *Proc. Nat. Acad. Sci. U.S.A. 70*, 183-186.

22. Goldner, H., Girardi, A. J., Larson, V. M., Hilleman, M. R. (1964). Interruption of SV40 virus tumorigenesis using irradiated homologous tumor antigen. *Proc. Soc. Exp. Biol. and Med. 117*, 851-857.

23. Grant, J. P. and Wells, S. A. (1974). *Surgery* (in press).

24. Grant, J., Ladisch, S., and Wells, S. A. (1974). Immunologic similarities between fetal cell antigens and tumor cell antigens in guinea pigs. *Cancer 33*, 376-383.

25. Hannon, W. H., Anderson, N. G., and Coggin, J. H. (1974). The relationship of sialic acid to the expression of fetal antigens in the developing hamster fetus. *World Symposium on Chemical Carcinogenesis.* J. Di Paolo and P. T'so, eds., Academic Press, New York pp. 669-684.

26. Haughton, G. and Nash, D. R. (1969). Transplantation antigens and viral carcinogenesis. *Prog. Med. Virol. 11*, 248-306.

27. Hellstrom, I., Hellstrom, K. E., Sjogren, H. O., and Warner, G. A. (1971). Demonstration of cell-mediated immunity to human neoplasms of various histological types. *Int. J. Cancer 7*, 1-16.

28. Herberman, R. B., Ting, C. C. and Lavrin, D. H. (1971). Immune reactions to virus-induced leukemia in animals immunized with fetal tissues. *Proc. of the First Conference and Workshop on Embryonic and Fetal Antigens in Cancer 1*, 259-266.

29. Kit, S., Kurimura, T., and Dubbs, D. R. (1969). Transplantable mouse tumor line induced by injection of SV40-transformed mouse kidney cells. *Int. J. Cancer 4*, 384-392.

30. Klein, G. Sjogren, H. O., Klein, E., Hellstrom, K. E. (1960). Demonstration of resistance against methylcholanthrene-induced sarcomas in the primary autochthonous host. *Cancer Res. 20*, 1561-1576.

31. Klein, G. (1966). Tumor Antigens. *Ann. Rev. Microbiol. 20*, 223-252.

32. LeMevel, B. P., and Wells, S. A. (1973). Foetal antigens cross-reactive with tumor-specific transplantation antigens. *Nature (London) 224*, 183-184.

33. Morton, D. L., Eilber, F. R., Joseph, S. L., Wood, W. C., Trahan, E., and Ketcham, A. S. (1970). Immunological factors in human sarcomas and melanomas: A rational basis for immunotherapy. *Ann. of Surgery 172*, 740-749.

34. Mukherji, B., and Hirshaut, Y. (1973). Evidence for fetal antigen in human sarcoma. *Science 181*, 440-442.

35. Old, L. J. and Boyse, E. A. (1964). Immunology of experimental tumors. *Ann. Rev. Med. 15* 167-186.

36. Old, L. J., and Boyse, E. A., Geering, G., and Oettgen, H. F. (1968). Serological approaches to the study of cancer in animals and man. *Cancer Res. 28*, 1288-1299.

37. Pearson, G. and Freeman, G. (1968). Evidence suggesting a relationship between polyoma virus-induced transplantation and normal embryonic antigens. *Cancer Res. 28*, 1665-1673.

38. Prehn, R. T. (1967). The significance of tumor-distinctive histocompatibility antigens. In *Cross-Reacting Antigens and Neoantigens*. ed. J. J. Trentin. Williams and Williams Co., Baltimore. 105-117.

39. Schone, G. (1906). *Munchener Medizinische Wochenschrift 51*, 1.

40. Stonehill, E. H. and Bendich, A. (1970). Retrogenic expression: The reappearance of embryonal antigens in cancer cells. *Nature (London) 228*, 370-372.

41. Ting, C. C., Herberman, R. B., Lavrin, D. H., and Shiu, G. (1971). Tumor-specific cell surface antigens in papova-virus-induced tumors and their relationship to fetal antigens in *Embryonic and Fetal Antigens in Cancer 1*, U. S. Dept. Commerce, Springfield, Va. 223-235.

42. Ting, C. C., Lavrin, D. H., Herberman, R. B. (1972). Antibodies to fetal antigens. In *Embryonic and Fetal Antigens in Cancer 2*, U. S. Dept. of Commerce, Springfield, Va. 111-114.

43. Ting, R. C. (1968). Failure to induce transplantation resistance against polyoma tumor cells with syngeneic embryonic tissues. *Nature 217*, 858-859.

44. Winslow, S. G. (1973). Unresponsiveness of the female syrian golden hamster to fetal immunization as related to female hormone levels and parity. Thesis University of Tennessee, Knoxville.

45. Zamcheck, N., Moore, T. L., Dhar, P. Kupchik, H. Z. (1972). Summary of current status (March 1972) of clinical studies of carcinoembryonic antigen (CEA). In *Embryonic and Fetal Antigens in Cancer. 2*, U. S. Dept. of Commerce, Springfield, Va. 209-214.

SERUM FACTORS AND CELL-MEDIATED DESTRUCTION OF TUMOR CELLS

Sylvia Pollack

Department of Microbiology
University of Washington Medical School
Seattle, Washington 98195

Introduction

During the course of tumor development, the host generates a complex of specific immune responses to the neoplastic cells. As monitored *in vitro*, the reaction includes both humoral and cell-mediated components which vary in their time of appearance and effects upon tumor cells. I will briefly consider a number of these responses (listed in Table I) with the main emphasis on two serum effects, arming of nonsensitized lymphoid cells and blocking of the cytotoxicity produced by sensitized lymphocytes.

Methods

Many of the conclusions to be discussed here are based on data obtained with an *in vitro* microcytotoxicity test (Takasugi and Klein, 1970; Hellström *et al.*, 1971a). In this assay, tumor cells which have been grown in monolayers in tissue culture, are seeded into the 96 wells of a Falcon Microtiter II plastic tissue culture plate. The cells are allowed to attach to the bottoms of the wells and then are treated sequentially with serum and lymphocytes. After a 30-40 hour incubation the surviving cells are stained and counted visually.

To calculate the arming effect, a comparison is made between the number of target cells surviving after treatment with normal (i.e., non-immune) serum and normal lymphocytes, compared to the number surviving after treatment with the test serum and normal lymphocytes. If the serum has an arming effect, there will be significantly fewer cells surviving after treatment with the test serum and normal lymphocytes than with the normal-normal control.

To determine the blocking effect of a serum, a comparison is made between the amount of cell-mediated cytotoxicity produced by immune lymphocytes in the presence of normal serum with that produced in the

187

TABLE 1.

Immune responses which may impede or favor tumor development

CON	PRO
Cell mediated cytotoxicity (CMC)	Blocking
By sensitized lymphoid cells	
By armed lymphoid cells	Immunostimulation
Potentiation of CMC	
Unblocking of CMC	
C' dependent cytotoxic antibody	

*As monitored *in vitro*

presence of immune or test serum. When a test serum has a blocking effect, the cellular immunity detected in the presence of normal serum will be no longer detectable or only partially detectable.

Immune Factors Impeding Tumor Growth

a) Cell mediated cytotoxicity

It is well established that most, if not all, tumors possess tumor-specific transplantation antigens (TSTA) in animals (Old and Boyse, 1964; Prehn and Main, 1957; Sjögren, 1965) and in man (Hellström *et al.*, 1971a). The antigenicity of animal tumors was detected by the demonstration of specific resistance to rechallenge after removal of the primary tumor. It has been further shown by tumor neutralization tests in which tumor cells are mixed with sensitized or control lymphoid cells and injected into appropriate recipients (Klein *et al.*, 1960; Old and Boyse, 1964), and by *in vitro* colony inhibition or microcytotoxicity tests (Hellström, 1965; Hellström and Hellström, 1971) that lymphoid cells of sensitized donors are specifically cytotoxic to the immunizing tumor.

Presumably, much of the lymphoid cell-mediated cytotoxicity is a function of sensitized T (thymus-derived) lymphocytes (see review of T cell activities by Raff elsewhere in this volume). Recent evidence, however, suggests that another mechanism may also be operative in lymphoid cell-mediated anti-tumor immunity. Some of these observations will be discussed in detail here.

b) Arming: Serum-mediated lymphoid cell-dependent cytotoxicity

1) The studies to be discussed grew out of the observation that sera from tumor-bearing mice were specifically cytotoxic to tumor cells *in vitro* when normal unstimulated lymph node cells were added (Pollack *et al.*, 1972). This cytotoxic effect (a) requires lymphoid cells, i.e. the sera by themselves are not

cytotoxic, (b) is immunologically specific, and (c) functions in syngeneic systems in contrast to an early report on a similar lymphoid cell-dependent phenomenon (Möller, 1965) (d) The classical complement cascade is not required for the reaction to proceed (Pollack and Nelson, 1973b).

Serum-mediated lymphoid cell-dependent cytotoxicity can be demonstrated against both primary virus-induced sarcomas and syngeneic transplantable methylcholanthrene (MCA)-induced sarcomas (Pollack et al., 1972). Similar findings have recently been reported for other syngeneic tumor systems by Kiesling and Klein (1973), de Landazuri (1973) and for human tumors (Hellström et al., 1973b).

Typical of the results obtained in the mouse tumor systems are the data in Table II on the survival of two different mouse tumor target cell lines after treatment with one of three different sera plus normal mouse lymph node cells. The lymphoid cell-dependent cytotoxicity was specific for the particular tumor against which the anti-serum was raised. No specific cytotoxicity was seen in the absence of lymphoid cells.

In many aspects, this tumor-specific cytotoxicity appears similar to the lymphocyte-dependent antibody-mediated killing of ^{51}Cr labeled target cells originally described in heterologous systems by Perlmann and Holm (1968),

TABLE 2.
Specificity of lymphoid cell-dependent cytotoxicity.

Target	Serum	Normal Lymph Node Cells	Cells/well ± S.E.	% Cytotoxicity[+]
MCA 1089	Normal	+	80.4 ± 3.3	
	Anti-MOL	+	89.0 ± 5.6	31.2%
	Anti-MCA	+	55.3 ± 4.2	*
	Normal	−	71.0 ± 3.0	
	Anti-MOL	−	73.9 ± 6.9	
	Anti-MCA	−	90.8 ± 10.7	
MOLONEY	Normal	+	39.0 ± 2.7	
	Anti-MOL	+	20.8 ± 1.9	46.7%
	Anti-MCA	+	46.8 ± 4.1	*
	Normal	−	33.3 ± 3.8	
	Anti-MOL	−	33.8 ± 3.4	
	Anti-MCA	−	42.6 ±4.4	

*p<0.001

[+]% cytotoxicity + [(NS + NL) − (IS + NL)]/(NS + NL) x 100 where (NS + NL) is the number of target cells surviving in the presence of normal serum and normal lymph node cells; (IS + NL) is the number of cells in the presence of immune serum and normal LNC.

MacLennan and Lewis (1968), and by MacLennan and Harding (1970). Typically, in those experiments, chicken erythrocytes or human Chang cells prelabeled with ^{51}Cr were treated with a heterologous antiserum and heterologous or autologous unstimulated lymphoid cells. Controls were treated with normal serum and lymphoid cells. Specific release of ^{51}Cr from the target cells could be detected. These studies have been comprehensively reviewed by Perlmann and Holm (1969). Recently, similar results have been obtained in other heterologous systems (e.g. Möller and Fvehag, 1972) and in human systems where the serum factor has been called lymphocyte-dependent antibody (LDA) (e.g., Yust et al., 1973).

Specific lymphoid cell-dependent cytotoxicity to tumor cells can be demonstrated even at very high serum dilutions (Pollack and Nelson, 1973b). Complement-dependent and lymphoid cell-dependent cytotoxicity produced by a serum taken from a mouse which carried a large Moloney virus-induced sarcoma is illustrated in Figure 1. Fresh rabbit serum was used as a source of complement. Although the complement-dependent cytotoxicity was not significant at serum dilutions greater than 1:320, the lymphoid cell dependent cytotoxicity was significant even at the highest dilution tested, 1:2560. In other experiments, the end point for lymphoid cell-dependent cytotoxicity is generally about $1:10^5$ (unpublished observations).

This remarkable activity is similar to the degree of activity reported by the Perlmann and MacLennan groups in their studies of lymphocyte-dependent cytolysis of antibody-coated target cells (Holms and Perlmann, 1969; MacLennan et al., 1970). In their heterologous systems, antisera diluted $1:10^6$ will produce lymphoid cell-dependent cytotoxicity. Perhaps the surprising fact is that analogous effects are seen at high dilutions in syngeneic tumor systems.

A repeatable feature of lymphoid cell dependent cytotoxicity to tumor cells as assayed in the microcytotoxicity test is that at an intermediate range of dilutions, e.g. 1:20 to 1:100, cytotoxicity cannot be detected (cf. Figure 1). A similar effect has been seen in other studies using ^{51}Cr labelled target cells (Koren and Trefts, 1973) and has been likened to a prozone effect.

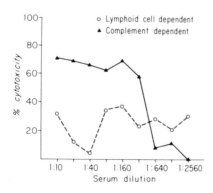

Fig. 1. *Comparison of the cytotoxicity produced on Moloney sarcoma target cells be sera from mice bearing progressively growing Moloney virus-induced sarcomas and either fresh rabbit serum as a complement source (Δ-Δ complement-dependent cytotoxicity) or non-sensitized syngeneic lymph node cells (o—o lymphoid cell dependent cytotoxicity).*

However, in the microcytotoxicity test where both cell death and cell proliferation can be monitored, stimulation of target cell growth is often observed (Pollack and Nelson, 1973b; unpublished observations). How and if this stimulation of cell proliferation relates to immunostimulation (Prehn and Lappe, 1971; Prehn, 1972; see below), has yet to be determined.

The identity of the serum factor which mediates lymphoid cell dependent cytotoxicity to tumor cells has not yet been established. In heterologous test systems, the data indicate that 7s immunoglobulins and antigen-antibody complexes are involved (MacLennan et al., 1969; Perlman et al., 1972).

The effector cell for serum-mediated lymphoid cell dependent cytotoxicity in syngeneic tumor systems also has not been definitely identified. In the heterologous test systems, the effector cells have the appearance of lymphocytes in electron microscope preparations (Biberfeld and Perlman, 1970).

In all systems investigated, the serum-dependent cytotoxic effect does not require T cells. Elimination from the effector cell population of theta-bearing cells by treatment with anti-theta serum and complement does not diminish the cytotoxicity (van Boxel et al., 1972). Spleen cells of thymus-deprived (i.e. thymectomized, lethally irradiated and bone marrow-reconstituted mice) are as effective as those from normal mice (van Boxel et al., 1972). Similarly, lymphoid cells from thymus-deprived rats are as effective as those from normal rats (Harding et al., 1971). Thymus cells (mainly T cells) are not efficient effector cells (Perlmann and Holm, 1969) while spleen cell preparations are (Perlmann et al., 1970). Spleen cells from congenitally thymusless (nude) mice are more active on a cell for cell basis than those from the spleen of a normal mouse (Kiesling and Klein, 1973).

Lymphoid-cell-dependent cytotoxicity can be diminished by treatment of the effector cell population with anti-lymphocyte serum and complement (Holm and Perlmann, 1969) or by anti-kappa chain serum and complement (van Boxel et al., 1972), suggesting that B cells are involved. However, further studies have implicated a subclass of B cells which bear a receptor for complement (van Boxel, et al., 1973) or an as yet unidentified cell type with receptors for the Fc region of antibody (Möller and Fvehag, 1972). Greenberg et al., (1973) have suggested that the effector cell is a non-T, non-Ig bearing monocyte.

2) In the experiments presented above on lymphoid cell dependent cytotoxicity to tumor cells, the serum was added to the target tumor cells prior to addition of the lymphocytes. In an attempt to directly arm the lymphocytes, 0.1-0.5 ml sera from tumor bearing mice were injected intraperitioneally to nonimmunized syngeneic mice. Four hours later the lymph nodes were excised, a single cell suspension prepared and the effect of those cells assayed on tumor target cells. The lymph node cells could be shown to be specifically cytotoxic to the tumor against which the serum was directed (Pollack, 1973). The results from one such experiment, which illustrate specificity in a reciprocal criss-cross experiment are shown in Table 3.

TABLE 3.
Specific in vitro *cytotoxicity produced by non-immune lymph node cells passively armed* in vivo.

LNC donor injected with serum	Target cells	No. target cells/well	% reduction[1]	p
0.5 ml normal		34.4 ± 2.8		
	MOL-1165		29.0	0.01
0.5 ml day 10 after MSV		24.4 ± 2.1		
			25.1	0.05
0.5 ml day 10 after 1085		32.6 ± 3.1		
0.5 ml normal		78.4 ± 3.1		
	MCA-1085		22.8	0.0005
0.5 ml day 10 after MSV		78.1 ± 4.6		
			22.5	0.005
0.5 ml day 10 after 1085		60.5 ± 2.8		

[1] Reduction in number of surviving tumor cells in the presence of lymph-node cells (LNC) from a mouse injected 4 hours previously with sera from mice bearing that tumor compared to number of surviving cells in the presence of LNC from mice pretreated with normal serum or sera from mice bearing a different tumor.

On the basis of such experiments, we have called the serum factor a lymphoid arming factor. It appears to be able to bind to either the target cell or the lymphocytes first and then couples together the target and effector cells, making possible some ensuing cytotoxic reaction. This mechanism could conceivably provide a versatile defense *in vivo* since the arming factor would be carried about by circulating lymphocytes or could attach to tumor cells, utilizing local lymphoid cells as effectors. Moreover, since the arming factor is active at high dilutions, the products (Ig?) of a few cells could arm many killer cells, greatly amplifying a cytotoxic response.

3) To look at the relationship between arming activity in the serum and tumor growth, we chose an experimental model where the tumor is rejected by the host's own immunologic defenses, Moloney virus-induced sarcomas in BALB/C mice (Fefer *et al.*, 1967, 1968). Tumors appear 5 days after an intramuscular injection of Moloney sarcoma virus (MSV), grow for approximately one week, then are rejected. The tumor regresses completely by about 3-4 weeks after virus injection if the recipient mouse is immunocompetent. The tumors progress and kill in immunologically suppressed or incompetent mice.

Adult BALB/C mice were injected with MSV and bled at various intervals so that sera were obtained from day 2 until day 45 after virus injection. High levels of arming activity were found during the period of tumor regression (unpublished observations). However, since the cytotoxic antibody blocking, unblocking, and potentiating ability of those sera were not tested concurrently, the contribution of arming to tumor rejection could not be inferred.

The most dramatic results obtained in these studies were at very early times after injection of MSV, during tumor induction and early growth, when immune responses are first called into play.

The arming activity (i.e. in this case, the ability to induce lymphoid cell dependent cytotoxicity) of 14 different sera collected 48 hours after injection of MSV is shown in Figure 2. Eight of 14 sera produced significant lymphoid cell dependent cytotoxicity. The mean cytotoxicity produced by all 14 samples was 25%.

Since it was possible that the arming activity had been passively transferred with the crude Moloney virus preparation, some mice were bled 4 hours after an intraperitoneal injection of MSV and their sera tested for arming activity. As illustrated by the example in Table 4, no arming was detected at a 1:10 dilution.

Several early sera were tested concurrently for arming and blocking activities. As shown in Table 5, sera taken 1 day after injection of MSV can induce lymphoid cell dependent cytotoxicity by non-immune lymphoid cells but do not block cytotoxicity produced by immune lymphocytes.

The early arming response may be a general feature of the immune response to particulate antigens since we have detected specific arming activity in sera of mice 1-2 days after immunization with MSV, Moloney sarcoma cells, methylcholanthrene-induced myosarcoma or bladder sarcoma cells, and allogeneic spleen cells (unpublished observations).

c) Early cellular immunity

Given the facts that lymph node cells can be armed *in vivo* by passively transferred arming serum (Pollack, 1973) and that arming factor exists in the serum of an animal 1-2 days after immunization (Pollack and Nelson, 1973a and above), an obvious prediction is that an animal will arm its own cells and that cellular immunity (cytotoxic lymphocytes) should be detectable at early times after immunization.

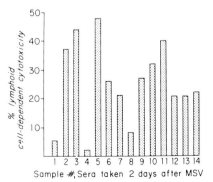

Fig. 2. *Amount of lymphoid cell-dependent cytotoxicity produced by 14 different sera taken 2 days after injection of Moloney sarcoma virus.*

TABLE 4.
Failure of serum taken 4 hours after injection of MSV to induce lymphoid cell-dependent cytotoxicity.

Target	Serum (1:10)*	LNC	# cells/well ± S.E.
NOL-1175	MSV	+	35.1 ± 3.1
	PBS	+	32.5 ± 2.7
	MSV	−	37.5 ± 4.6
	PBS	−	35.5 ± 5.1

*Serum obtained 4 hours after an intraperitoneal injection of a 1:7 dilution of stock Moloney virus preparation (MSV) or after an equal volume of saline (PBS).

We have looked for specific cytotoxicity by lymph node cells taken 24-48 hours after injection of Moloney virus. The cellular cytotoxicity obtained with different doses of lymph node cells from 3 different mice tested 48 hours after injection of MSV is shown in Figure 3. At the higher effector cell: target cell ratios tested, the lymph node cells were specifically cytotoxic. Normal syngeneic brain cells were used as control target cells and the data presented here were corrected for a low level of non-tumor specific cytotoxicity.

Similar results have been obtained with lymphoid cells tested at early times after transplantation of syngeneic MCA sarcomas. In general, the cell

TABLE 5.
Arming and blocking activities of sera taken a day after injection of Moloney sarcoma virus.

Serum (1:6)	LNC	# cells/well ± S.E.
Normal	Normal	40.7 ± 3.4
	Immune	24.8 ± 1.5
Day 1 (♀)	Normal	27.0 ± 1.7
	Immune	25.0 ± 2.7
Day 1 (♂)	Normal	26.8 ± 2.6
	Immune	21.7 ± 2.6

Immune cells:	39.1% decrease	$p < 0.002$
Arming (♀):	33.6% decrease	$p < 0.005$
Arming (♂):	34.1% decrease	$p < 0.01$
Blocking (♀):	0.5% abrogation	
Blocking (♂):	−19.2% abrogation	

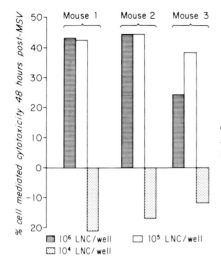

Fig. 3. *Cell mediated cytotoxicity produced by lymph node cells (LNC) of BALB/c mice injected 2 days previously with Moloney sarcoma virus.*

mediated cytotoxicity detected at 24 hours is weaker at a given LNC dose and less consistently detectable than that found at 48 hours (e.g., significant cytotoxicity at 10^5 LNC/well produced by 2/6 tested at 24 hours vs. 6/7 tested at 48 hours).

Early cell mediated immunity to tumors has recently been reported in several systems. Using tumor allografts, Forman and Britton (1973) have detected lymphocyte mediated cytotoxicity 3 days after immunization. In syngeneic tumor systems, Bansal and Sjögren (1973) found cellular immunity to polyoma tumors at 3 days and Lamon *et al.* (1972) found cellular immunity to Moloney sarcomas at 3 days.

Thus, there is no question that cytotoxic lymphoid cells can be demonstrated at early times after challenge with a transplantable tumor or a tumor-inducing virus. The simultaneous early appearance of cellular immunity and arming factor suggests that arming of lymphoid cells may be a normal biological response to antigenic challenge.

d) Cytotoxic antibody

Complement dependent cytotoxic antibody can be demonstrated in many tumor systems (e.g. Slettenmark and Klein, 1962; Old *et al.*, 1963; Hellström, 1965; Lewis *et al.*, 1969) although the use of heterologous complement in these assays makes extrapolation to the *in vivo* situation somewhat suspect. Some antibody may produce cytostasis rather than cytolysis *in vivo* (Ambrose *et al.*, 1971): i.e., tumor cells implanted in cell impermeable chambers in the peritoneum of immunized mice can remain viable without increase or decrease in number.

e) Potentiation

Recently it has been noted that some sera from tumor-bearing patients, rats or quail can potentiate or increase the specific cytotoxicity produced by immune lymphocytes (Hellström *et al.*, 1971, 1973; Skurzak and Klein, 1972; Hayami *et al.*, 1973). The mechanism of potentiation is not known but at least part of the activity may be related to arming of effector cells.

f) Unblocking

Another serum factor has been described which can free cell-mediated immunity from the restraints put upon it by blocking factors (which are discussed in detail below).

This unblocking factor has been demonstrated *in vitro* by mixing serum known to have blocking activity with serum from an animal or patient whose tumor (of the same type) has regressed or been removed. The blocking serum prevents the expression of cytotoxicity by immune lymphoid cells. But in the presence of a mixture of blocking serum plus unblocking serum, cellular immunity is expressed and the target cells are killed. Unblocking activity, which is tumor-specific, has been demonstrated with mouse Moloney sarcomas (Hellström and Hellström, 1970), rat polyoma tumors (Bansal and Sjögren, 1971), quail Rous-virus-induced sarcomas (Hayami *et al.*, 1972), and human melanomas and renal carcinomas (Hellström *et al.*, 1972). Although the identity and mode of action of the unblocking factor have not been definitely established, available data indicate that it may be an antibody, since it can be absorbed out with the respective tumor cells and which may act by binding to the antigen parts of blocking factors (see below).

Bansal and Sjögren (1971, 1972) have reported some success in treating polyoma tumors *in vivo* in rats with large doses of serum known to have unblocking activity *in vitro*. Recipients of unblocking serum had smaller tumors and survived longer than recipients of control serum, and in some rats treated with unblocking serum the tumors regressed. Although *in vitro* studies on the treated rats showed that rats receiving unblocking serum had lower blocking titers and suggested that the unblocking factors may have exerted their effect this way, the possibility cannot be excluded that the *in vivo* effects were at least partially due to arming or potentiation.

Immune Factors Promoting Tumor Growth

An obvious question is how do tumors circumvent these various anti-tumor responses and continue to grow? Among the possible explanations are loss by the tumor of its tumor-specific antigens or coating of the tumor cell surface with sialomucin either of which could prevent recognition of the tumor cells by the immune system. Some tumor cell types may be capable of division at a rate greater than can be dealt with by the immune system. In addition, the *in situ* relationship of tumor cells and immune lymphoid cells may not favor tumor cell killing. There also is evidence for immune responses which actively thwart anti-tumor immunity.

a) Blocking

Serum factors which can block the cytotoxic effect of immune lymphocytes, allowing the tumor cells to survive, have been detected in many tumor systems and are extensively reviewed elsewhere (Hellström and Hellström, 1971, 1972; Heppner, 1972). Blocking of anti-tumor cellular immunity *in vitro* appears to be a correlate of the *in vivo* phenomenon of enhancement (Kaliss, 1958). Like cell-mediated immunity, blocking is immunologically specific, e.g., sera taken from a mouse bearing a progressively growing Moloney-virus induced sarcoma will not block the cytotoxic effect of cells immune to a methylcholanthrene-induced sarcoma on the MCA sarcoma cells.

In studies on polyoma virus-induced tumors in rats, Bansal and Sjogren (1972) demonstrated the following aspects of the relationships of cellular immunity and blocking factors to tumor course:

1) Cell mediated immunity was present during tumor growth and after tumor removal. In animals bearing very large tumor loads, cellular reactivity was reduced.

2) Blocking activity, as detected *in vitro*, was present during tumor growth and became undetectable after tumor removal.

3) Blocking serum could specifically enhance the growth of polyoma tumor cells transplanted to another rat.

4) Rechallenge of rats which had cytotoxic lymphocytes but no blocking activity in their sera did not grow while in rats with cellular immunity plus blocking activity, a challenge did grow.

These studies thus provide correlative *in vitro* and *in vivo* evidence for a role of blocking factor in the growth of tumors.

Evidence suggesting a correlation of blocking activity in the serum of human patients with their clinical status has emerged from sequential studies on melanoma patients (Hellström *et al.*, 1973b; Byrne *et al.*, 1973). In general, blocking activity was negligible while a patient was clinically tumor free and reappeared when metastases were noted. Although a rise or fall in blocking activity often paralleled the clinical course, at any particular test the detectable blocking was not necessarily predictive.

The identity of the blocking factor has not been conclusively established although there is suggestive evidence that antigen-antibody complexes are involved (Sjögren *et al.*, 1971). Other studies suggest, however, that either antigen (Brawn, 1971; Currie and Basham, 1972) or antibody alone (Baldwin *et al.*, 1973) can block cell mediated immunity. On the basis of studies with rat hepatomas Baldwin recently suggested antigen alone may be a blocking agent during early stages of tumor development while antigen-antibody complexes play a major role in later stages of tumor growth (Baldwin et al., 1973).

Further corroboration of the view that blocking may be mediated by different entities at different stages of the immune response came from studies on Rous virus-induced sarcomas on Japanese quail by Hayami et al., (1972, 1973, 1974). Most sera from birds with progressively growing tumors were blocking, similar to results found in other model systems. In addition, some regressor sera had blocking activity. Some of these could block at either the target cell or the effector cell level while others could block only when incubated with the target cells. These data are compatible with the interpretation that in the first group of sera blocking was due to antigen-antibody complexes while in the second group only antibody directed against the tumor antigens acted as a blocker. Further experiments in the Rous system (Hayami et al., 1974) have suggested that antibody directed against the tumor antigens may block by facilitating release of antigens from the tumor cell surface with the consequent formation of blocking complexes.

In the Rous sarcoma system in quails, blocking activity has been shown to be bursa dependent (Hayami et al., 1972), i.e., surgically bursectomized birds which have greatly reduced immunoglobulin levels in their sera have no blocking activity. The spleen apparently plays a role in the elaboration of blocking factor in mice. Splenectomized mice produce little if any blocking factor when bearing progressively growing MCA sarcomas while sham-operated littermates have high blocking levels (Hellström et al., 1971). Spleens from tumor immunized mice can produce blocking factor when cultured in vitro (Nelson and Pollack, unpublished observations). Less vigorous growth of tumors in splenectomized mice (Pollack, 1971) is also consistent with the suggestion that the spleen plays a role in the in vivo production of blocking factor.

b) Immunostimulation

A second mechanism for encouraging tumor growth during its early stages is the recently proposed phenomenon of immunostimulation (Prehn and Lappe, 1971). Prehn (1972) has presented in vivo evidence suggesting that small numbers of immune lymphocytes may be trophic for weakly antigenic tumors – tumors which are probably more representative of clinical tumors in human patients than the strongly antigenic tumors used in many animal model studies. Medina and Heppner (1973) have obtained data with in vitro tests which support the immunostimulation concept using a weakly immunogenic tumor, i.e., specifically immune lymphoid cells stimulated tumor cell growth. Some anti-sera at dilutions in the range of 1:50 to 1:100 are also specifically stimulatory to the appropriate tumor cell line when normal lymphoid cells are present (Pollack and Nelson, 1973b; unpublished observations). A role for such immunostimulatory mechanisms for tumor growth in vivo remains to be established.

Summary

Until quite recently it would have been reasonable to assume that cell-mediated anti-tumor immunity was a T cell response and was "good" for the organism, while production of antibody by B cells would lead to blocking and was therefore a "bad" response. However, as discussed above, recent data suggest the following: Not all cell-mediated anti-tumor immunity is dependent on T cells. A serum factor, perhaps antibody produced by B cells, can "arm" non-T effector cells. Moreover, the presence of cytotoxic lymphoid cells may stimulate rather than inhibit tumor growth under particular conditions. The blocking of anti-tumor activity by serum factors may be an unfortunate manifestation of a mechanism which is required for maintenance of active tolerance to self and for protection of allogenic fetuses (Hellström and Hellström, 1972; Wegmen et al., 1971; Hellström, et al., 1969).

On the basis of the information and hypotheses presented here, we can speculate that the following sequence of events might occur in vivo: When the host first becomes cognizant of neoplastic cells, it responds humorally, by producing arming factor and cytotoxic antibody. These responses may provide a first line of defense. Alternatively, armed cells might conceivably be trophic for the developing tumor. If the tumor escapes eradication by the early responses and continues to grow, antigen will be shed and further facets of the immune response elicited. Antigen-antibody complexes will form and may block the activity of immune (i.e. armed non-T and cytotoxic T) cells. If the cell mediated immunity is blocked and the tumor continues to grow, more tumor antigens will be released, putting the host into a state of antigen excess which can be expected to further block any cellular immunity. In other cases, however, cytotoxic effector cells in the presence of unblocking antibody may be able to seek out and destroy the tumor before it overwhelms the host.

Our understanding of the interrelationships of these humoral and cellular responses is rudimentary and our knowledge of how other factors, such as the genetic and hormonal make-up of the host, impinge on the anti-tumor response is practically nil. However, the goal of the studies discussed here is obvious: to increase our understanding to the point where the humoral and cell mediated responses to a tumor can be manipulated for effective immunotherapy and immunoprophylaxis.

ACKNOWLEDGEMENTS

I thank Karen Nelson for her many contributions to this work and Drs. Gloria Heppner, Karl Erik Hellström, William Elkins and Irwin Bernstein for their critical reading of the manuscript.

This work has been supported by the National Institutes of Health CA-14697 and by the following grants to Drs. Ingegerd and Karl Erik Hellström: American Cancer Society T-453, NIH CA-10188 (KEH) and CA-10189 (IH), and contract NIH-NCI-71-2171 within the Special Virus-Cancer Program.

REFERENCES

1. Ambrose, K. R., N. G. Anderson and J. H. Coggin, 1971. Cytostatic antibody and SV40 tumor immunity in hamsters. *Nature 233*:321-324.

2. Baldwin, R. W., M. J. Embleton and R. A. Robins, 1973. Cellular and humoral immunity to rat hepatoma-specific antigens correlated with tumor status. *Int. J. Cancer 11*:1-10.

3. Baldwin, R. W., M. R. Price and R. A. Robins, 1973. Inhibition of hepatoma-immune lymph node-cell cytotoxicity by tumor-bearer serum, and solubilized hepatoma antigens. *Int. J. Cancer 11*:527-535.

4. Bansal, S. C. and H. O. Sjögren, 1971. "Unblocking" serum activity *in vitro* in the polyoma system may correlate with antitumor effects of antiserum *in vivo*. *Nature 233*:76-78.

5. Bansal, S. C. and H. O. Sjögren, 1972. Counteraction of the blocking of cell-mediated tumor immunity by inoculation of unblocking sera and splenectomy: immunotherapeutic effects on primary polyoma tumors in rats. *Int. J. Cancer 9*:490-509.

6. Bansal, S. C. and H. O. Sjögren, 1973. Correlation between changes in antitumor immune parameters and tumor growth *in vivo* in rats. *Fed. Proc. 32*:165-172.

7. Biberfeld, P., and P. Perlmann, 1970. Morphological observations on the cytotoxicity of human blood lymphocytes for antibody-coated chicken erythrocytes. *Exp. Cell. Res. 62*:433-440.

8. Brawn, R. J., 1971. *In vitro* desensitization of sensitized murine lymphocytes by a serum factor (soluble antigen?). *Proc. Nat. Acad. Sci. 68*:1634-1638.

9. Currie, G. A. and C. Basham, 1972. Serum mediated inhibition of the immunological reactions of the patient to his own tumor: a possible role for circulating antigen. *Brit. J. Cancer 26*:427-438.

10. deLandazuri, M. O., 1973. Synergistic cytotoxic response involving lymphocyte mediated antibodies to a syngeneic lymphoma. *Fed. Proc. 32*:998 abs.

11. Fakhiri, O. and J. R. Hobbs, 1972. Target cell death without added complement after cooperation of 7s antibodies with non-immune lymphocytes. *Nature New Biol. 235*:177-178.

12. Fefer, A., J. L. McCoy and J. P. Glynn, 1967. Induction and regression of primary Moloney sarcoma virus-induced tumors in mice. *Cancer Res. 27*:1626-1631.

13. Fefer, A., J. L. McCoy, K. Perk and J. P. Glynn, 1968. Immunologic, virologic and pathologic studies of regression of autochthonous Moloney sarcoma virus-induced tumors in mice. *Cancer Res. 28*:1577-1585.

14. Greenberg, A. H., L. Hudson, L. Shen and I. M. Reitt, 1973. Antibody-dependent cell-mediated cytotoxicity due to a "null" lymphoid cell. *Nature New Biol. 242*:111-113.

15. Halliday, W. J., 1971. Blocking effect of serum from tumor-bearing animals on macrophage migration inhibition with tumor antigens. *J. Immunol. 106*:855-857.

16. Harding, B., D. J. Pudifin, F. Gotch and I. C. M. MacLennan, 1971. Cytotoxic lymphocytes from rats depleted of thymus processed cells. *Nature New Biol. 232*:80-81.

17. Hayami, M., I. Hellström, K. E. Hellström and K. Yamanouchi, 1972. Cell-mediated destruction of Rous Sarcomas in Japanese quails. *Int. J. Cancer 10*:507-517.

18. Hayami, M., I. Hellström and K. E. Hellström, 1973. Serum effects on cell-mediated destruction of Rous Sarcomas. *Int. J. Cancer, 12*:667-688.

19. Hayami, M., I. Hellström, K. E. Hellström and D. R. Lannin, 1974. Further studies on the ability of regressor sera to block cell-mediated destruction of Rous Sarcomas. *Int. J. Cancer 13*:43-53.

20. Hellström, I. and K. E. Hellström, 1969. Studies on cellular immunity and its serum-mediated inhibition in Moloney virus-induced mouse sarcomas. *Int. J. Cancer* 4:587-600.

21. Hellström, I. and K. E. Hellström, 1970. Colony inhibition studies on blocking and non-blocking serum effects on cellular immunity to Moloney sarcomas. *Int. J. Cancer* 5:195-201.

22. Hellström, I. and K. E. Hellström, 1972. Can "blocking" serum factors protect against autoimmunity. *Nature* 240:471-473.

23. Hellström, I., K. E. Hellström and H. O. Sjögren, 1970. Serum mediated inhibition of cellular immunity to methylcholanthrene-induced murine sarcomas. *Cell. Immunol.* 1:18-30.

24. Hellström, I., K. E. Hellström, H. O. Sjögren and G. A. Warner, 1971a. Demonstration of cell-mediated immunity to human neoplasms of various histological types. *Int. J. Cancer* 7:1-16.

25. Hellström, I., K. E. Hellström, H. O. Sjögren and G. A. Warner, 1971b. Serum factors in tumor-free patients cancelling the blocking of cell-mediated tumor immunity. *Int. J. Cancer* 8:185-191.

26. Hellström, I., K. E. Hellström and J. J. Trentin, 1973a. Cellular immunity and blocking serum activity in chimeric mice. *Cell. Immunol.* 7:73-84.

27. Hellström, I., K. E. Hellström and G. A. Warner, 1973b. Increase of lymphocyte-mediated tumor cell destruction by certain patient sera. *Int. J. Cancer, 12*:348-352.

28. Hellström, K. E. and I. Hellström, 1970. Immunological enhancement as studied by cell culture techniques. *Ann. Rev. Microbiol.* 24:373-398.

29. Hellström, K. E., I. Hellström and R. J. Brawn, 1969. Abrogation of cellular immunity to antigenically foreign mouse embryonic cells by a serum factor. *Nature* 224:914-915.

30. Heppner, G. H., 1969. Studies on serum-mediated inhibition of cellular immunity to spontaneous mouse mammary tumors. *Int. J. Cancer* 4:600-615.

31. Heppner, G., 1972. Blocking antibodies and enhancement. *Ser. Haemat.* 5:41-66.

32. Holm, G. and P. Perlmann, 1969. Inhibition of cytotoxic lymphocytes by antilymphocyte serum. *Transpl. Proc.* 1:420-423.

33. Kaliss, N., 1958. Immunological enhancement of tumor homografts: a review. *Cancer Res.* 18:992-1003.

34. Klein, G., H. O. Sjögren, E. Klein and K. E. Hellström, 1960. Demonstration of resistance against methylcholanthrene-induced sarcomas in the primary autochthonous host. *Cancer Res.* 20:1561-1572.

35. Koren, H. S. and P. Trefts, 1973. Antibody-dependent cell-mediated cytotoxicity to TNP-modified target cells in mice. *Fed. Proc.* 32:998 Abs.

36. Lamon, E. W., H. M. Skurzak and E. Klein, 1972. The lymphocyte response to a primary viral neoplasm (MSV) through its entire course in Balb/c mice. *Int. J. Cancer* 10:581-588.

37. Lewis, M. G., R. L. Ikonopesov, R. C. Nairn, T. M. Phillips, G. Hamilton-Fairley, D. C. Bodenham and P. Alexander, 1969. Tumor-specific antibodies in human malignant melanoma and their relationship to the extent of the disease. *Brit. Med. J.* 3:547-552.

38. MacLennan, I. C. M. and B. Harding, 1970. Failure of certain cytotoxic lymphocytes to respond mitotically to phytohaemagglutinin. *Nature* 227:1246-1248.

39. MacLennan, I. C. M. and G. Loewi, 1968. The effect of specific antibody to target cells on their specific and non-specific interactions with lymphocytes. *Nature* 219:1069-1071.

40. MacLennan, I. C. M., G. Loewi and B. Harding, 1970. The role of immunoglobulins in lymphocyte-mediated cell damage *in vitro*. I. Comparison of the effects of

target cell specific antibody and normal serum factors on cellular damage by immune and non-immune lymphocytes. *Immunology 18*:397-404.

41. MacLennan, I. C. M., G. Loewi and A. Howard, 1969. A human serum immunoglobulin with specificity for certain homologous target cells, which induced target cell damage by normal human lymphocytes. *Immunology 17*:897-910.

42. Medina, D. and G. Heppner, 1973. Cell-mediated "immunostimulation" induced by mammary tumor virus-free BALB/c mammary tumors. *Nature 242*:329-330.

43. Möller, E., 1965. Contact-induced cytotoxicity by lymphoid cells containing foreign isoantigens. *Science 147*:973-879.

44. Moller, G. and F. Fvehag, 1972. Specificity of lymphocyte-mediated cytotoxicity induced by *in vitro* antibody-coated target cells. *Cellular Immunol. 4*:1-19.

45. Old, L. J. and E. Boyse, 1964. Immunology of experimental tumors. *Ann. Rev. Med. 15*:167-186.

46. Old, L. J., E. A. Boyse and F. Lilly, 1963. Formation of cytotoxic antibody against leukemias induced by Friend virus. *Cancer Res. 23*:1063-1068.

47. Perlmann, P. and G. Holm, 1969. Cytotoxic effects of lymphoid cells *in vitro*. In: F. J. Dixon, Jr. and H. G. Kunkel (ed.). *Advances in immunology. 11*:117-195, Academic Press, New York.

48. Perlmann, P., H. Perlmann and P. Biberfeld, 1972. Specifically cytotoxic lymphocytes produced by preincubation with antibody-complexed target cells. *J. Immunol. 108*:558-561.

49. Pollack, S., 1973. Specific "arming" of normal lymph node cells by sera from tumor bearing mice. *Int. J. Cancer 11*:138-142.

50. Pollack, S., G. Heppner, R. J. Brawn and K. Nelson, 1972. Specific killing of tumor cells *in vitro* in the presence of normal lymphoid cells and sera from hosts immune to the tumor antigens. *Int. J. Cancer 9*:316-323.

51. Pollack, S. and K. Nelson, 1973a. Early appearance of "arming activity" after tumor induction: A first line of defense? *Fed. Proc. 32*:1015 Abs.

52. Pollack, S. and K. Nelson, 1973b. Effects of carrageenan and high serum dilution on synergistic cytotoxicity to tumor cells. *J. Immunol. 110*:1440-1443.

53. Prehn, R. T. and J. M. Main, 1957. Immunity to methylcholanthrene-induced sarcomas. *J. Nat. Cancer Inst. 18*:769-778.

54. Prehn, R. T., 1972. The immune reaction as a stimulation of tumor growth. *Science 176*:170-171.

55. Prehn, R. T. and M. A. Lappe, 1971. An immunostimulation theory of tumor development. *Transplantation Reviews 7*:26-54.

56. Sjögren, H. O., I. Hellstrom, S. C. Bansal and K. E. Hellstrom, 1971. Suggestive evidence that the "blocking antibodies" of tumor bearing individuals may be antigen-antibody complexes. *Proc. Nat. Acad. Sci. 68*:1372-1375.

57. Skurzak, H. M., E. Klein, T. O. Yoshida and E. W. Lamon, 1972. Synergistic or antagonistic effect of different antibody concentrations on *in vitro* lymphocyte cytotoxicity in the Moloney sarcoma virus system. *J. Exptl. Med. 135*:997-1002.

58. Slettenmark, B. and E. Klein, 1962. Cytotoxic and neutralization tests with serum and lymph node cells of isologous mice with induced resistance against Gross lymphomas. *Cancer Res. 22*:947-954.

59. Takasugi, M. and E. Klein, 1970. A microassay for cell-mediated immunity. *Transpl. 9*:219-227.

60. van Boxel, J. A., W. E. Paul, M. M. Frank and I. Green, 1973. Antibody-dependent lymphoid cell-mediated cytotoxicity: Role of lymphocytes bearing a receptor for complement. *J. Immunol. 110*:1027-1036.

61. van Boxel, J., J. D. Stobo, W. E. Paul and I. Green, 1972. Antibody-dependent lymphoid cell-mediated cytotoxicity: No requirement for thymus-derived lymphocytes. *Science* 175:194-196.

62. Wegmann, T. G., I. Hellström and K. E. Hellström, 1971. Immunological tolerance: "Forbidden clones" allowed in tetraparental mice. *Proc. Nat. Acad. Sci.* 68:1644-1647.

63. Yust, I., J. R. Wunderlich, D. L. Mann and D. N. Buell, 1973. Cytotoxicity mediated by human lymphocyte-dependent antibody in a rapid assay with adherent target cells. *J. Immunol.* 110:1672-1681.

LYMPHOCYTE DIFFERENTIATION AND IMMUNE SURVEILLANCE AGAINST CANCER*

G. J. V. Nossal

The Walter and Eliza Hall Institute of Medical Research
Melbourne, Victoria 3050, Australia

The last speaker in a Symposium including such a distinguished panel of experts has a difficult task, particularly when such a wide range of issues has been covered. However, I approach the challenge with much pleasure, because, in common with many of my immunological colleagues, I consider developmental biology as one of the last and most important frontiers. There are quite a few of us in cellular immunology who are really frustrated embryologists, and I must admit that those of my experiments which involve work with early embryos hold a special fascination for me. I think, also, that some of the operational features of immunology research, and in particular the fine single cell methods which exist, make the immune system bristle with models of interest to students of developmental biology.

In this lecture, I wish to cover two broad areas. First, I shall give a personal view of the present status of the immune surveillance hypothesis. Secondly, I will cover some of our more recent work at the Walter and Eliza Hall Institute on developmental aspects of lymphocyte formation and function, in particular as it relates to lymphocyte triggering. Clearly, mechanisms of lymphocyte activation are of key importance to an understanding of surveillance.

*Original work reported in this communication was supported by the National Health and Medical Research Council and the Australian Research Grants Committee, Canberra Australia; by the Volkswagen Foundation (Grant No. 11 2147); and by the United States Public Health Service (Grant AI 03958); and Contract No. NIH-NCI-G-72-3889 with the National Cancer Institute, National Institutes of Health, Department of Health, Education and Welfare, U.S.A.

This is Publication No. 1893 from the Walter and Eliza Hall Institute.

Immune Surveillance – Facts and Problems.

The underlying assumption of the immune surveillance notion is that a healthy lymphoid system constitutes a bulwark against development of malignancy. Lymphocytes are seen as suppressing many collections of altered, possibly premalignant cells arising in the body, so long as these display some surface component (or tumor-associated transplantation antigen, TATA) recognizable as foreign by autologous lymphocytes. A good example, would be cells infected by a potentially oncogenic virus. For example, human lymphocytes infected by the herpes-type Epstein-Barr virus behave in many respects like malignant cells. The blood picture and tissue pathology of a patient with acute infectious mononucleosis has features in common with that of acute lymphoblastic leukemia. It is possible that the virally-induced antigens which appear at the surface of the infected cells are instrumental in ensuring their immunological destruction, and that malignancy such as Burkitt's lymphoma, supervenes when *one* cell, for uncertain reasons, escapes this surveillance. By the same token, tumors induced by chemical means are usually antigenic in their hosts. The dominant antigens from the viewpoint of immune rejection are unique to each particular chemically induced tumor, though shared antigens can also be detected in some systems. A third group of tumor antigens exists, the so-called carcinoembryonic antigens (CEA). These are believed to reflect a reversion by the cancer cell to an embryonic pattern of synthesis. The CEA associated with human colonic and other cancers does not appear to cause an immune response in the host, and is therefore not of importance for immune surveillance. However, other types of carcinoembryonic antigens of more relevance to surveillance may exist. The classification of human tumor antigens does not exactly follow the classical picture derived from work with mice, and particular histologic types of cancer show considerable antigenic sharing.

Therefore, although problems of detail remain to be resolved, there is no dispute about the fact that most malignant tumors are antigenic in their host. It is also clear that in many cases the host mounts a substantial immune response against the tumor. Clearly, in the majority of cancers, this response is insufficient to halt the progress of the disease. The point at issue is whether these same considerations apply to premalignant cell collections present in the body, i.e. whether, and to what extent, an animal with a crippled immune system develops cancer more readily.

I wish to draw attention to five studies relevant to this problem. The first four were presented in detail at a recent major symposium (1, 4), whereas the fifth (5) is a study that is, as yet, unpublished, and I am grateful to my colleague Dr. Vernon Marshall for giving me permission to cite it. First, we owe to R. A. Good (1) an analysis of over 1,000 children afflicted with various forms of congenital defects of the immune apparatus. These included Bruton-type agammaglobulinemia, where the B lymphocyte system is seriously deficient; Wiscott-Aldrich syndrome, ataxia telangiectasia and other diseases affecting chiefly the T cell system; and the so-called "common variable" immunodeficiency. Almost ten per cent of these children developed a malignancy. This extremely high incidence in an age group where cancer is not common seems to give great support to immune surveillance. Secondly, Penn and others have analyzed cancer incidence in renal transplant recipients, there now being over 7,000 patients bearing kidney grafts for a year or longer, aided, of course, by drugs which inhibit the immune system. Of these, about one per cent have so far developed cancer. This is not a dramatically high number, but far greater than would be expected by chance.

While these two studies at first sight lend credence to the immune surveillance notion, a second look reveals an unexpected complexity. The distribution of types of malignancy occurring in the patients with depressed lymphocyte function is quite different from that occurring spontaneously in a general human population. In particular, there is an undue preponderance of lymphoid system malignancies – lymphomas, leukemias and reticuloses. This could be explained in various ways. Possibly such tumors are virally induced, and strongly antigenic; as such they may be able to emerge only in immune deficiency settings. Possibly the increased load of bacterial, viral and fungal antigens which the immunological cripple must bear *because* of his defenceless state leads to an over-stimulation of the lymphon, and thus to malignancy. Possibly the gene defects leading to immune deficiency predispose in a direct fashion to lymphoid malignancy, and the lympho-destructive drugs may do the same.

The third and fourth tests of the immune surveillance hypothesis involved a major literature search of experimental animal work on tumor induction and incidence in, respectively, immunosuppressed and immunostimulated animals. Interested readers are referred to Dr. O. Mäkelä's excellent paper (2). The studies gave general support to the surveillance notion, in that suppressed animals yielded more or earlier cancers, while animals whose lymphoid system had been non-specifically activated by adjuvant substances frequently gave fewer cancers. The results were more impressive for virus-induced than for chemically-induced or spontaneously-emerging tumors. A special note of caution was sounded about over-vigorous non-specific immunostimulation (3). Like all good things that are overdone, this can on occasion lead to disaster. Studies in mice have shown that the long term administration of zymosan, a stimulant of the RES, can result in an increase of spontaneous cancer.

The fifth example is again a human one, and its surprising feature is the high incidence of skin cancer in immunosuppressed patients. Dr. Vernon Marshall, Department of Surgery, University of Melbourne, has presented results derived from the Australasian Transplant Registry covering a ten year period from 1963 to 1973, during which time 1035 kidney grafts, mainly cadaver grafts, had been performed in Australasia. Over that period, 36 cases of cancer in transplant recipients had been reported, an incidence of 3%. Of these 6 were lymphomas, of which 5 proved fatal; 4 were visceral carcinomas, of which 2 proved fatal; 1 was a tumor in the transplanted kidney, which proved fatal; 1 was a pre-existent tumor which recurred after transplantation and proved fatal; and 24, or fully two-thirds of the series were cases of skin cancer, of which 2 proved fatal. Skin cancers thus occurred in 2% of the series of kidney transplantations.

A most unusual feature of the skin cancers occurring in these patients is that there were 20 squamous cell carcinomas and only 4 basal cell carcinomas, the ratio of 5 to 1 being an exact reversal of the normal ratio of 1 to 5 which is seen in the Australian population not on immunosuppressive therapy.

Comparison of these figures with world figures derived from the International Transplant Registry is of interest. In the world figures, which now embrace over seven thousand renal transplants, there is an overall incidence of about 1% of all tumors, but only one sixth of these are skin tumors.

In the Royal Melbourne Hospital's own series, there were 13 cases of skin cancer but the terrifying statistic was noted that amongst the patients surviving for more than 4 years, 17% have skin cancer and fully 30% have been noted to have multiple solar keratoses. Kerato-acanthomas, generally regarded as non-malignant conditions, have also been noted in some of these patients, and have been seen to undergo a particularly unfavourable course.

This remarkable increase in skin cancer in the Australian transplant patient as opposed to the world-wide transplant patient probably reflects the conjoint action of high exposure to ultraviolet radiation and predominantly fair skin on people that are being treated with drugs which are either antagonizing the normal immune surveillance function of the lymphocyte, or, conceivably which might be oncogenic.

We can therefore conclude that immune surveillance against oncogenically transformed cells is a reality, but of course it may not be the only important homeostatic mechanism. In fact, if it were one might have expected the cancer incidence in immunodeficiency states to be higher still. However, experimental and clinical examples of immunosuppression are rarely complete, as total immune deficiency is not compatible with life except in a germ-free isolator. Certainly, further studies on spontaneous cancer incidence in mice with various forms of congenital immunodeficiency should be performed with animals held long-term in the germ-free state. While contrary views are still current (6), my personal bias is to regard immune surveillance as a major force.

Lymphocyte Triggering and the Cancer Problem.

Next, I wish to consider what we know about lymphocyte activation that might be relevant to the immune surveillance problem, and to immune defence against cancer. Presumably, the growing premalignant clone could engage the attention of the lymphoid cells in one of three ways. First, as lymphocytes recirculate through tissues, there could be a direct physical encounter between the altered cell and the lymphocyte. This, though being a statistically unlikely event in view of the diversity of lymphocytes and the small size of the premalignant clone, would have one advantage from the viewpoint of immunocyte triggering, namely the probability that multivalent presentation of antigen to the lymphocyte would occur. It is thought that multivalent antigens can trigger lymphocytes more easily than univalent ones. Secondly, soluble tumor antigens could be shed into the extracellular fluid through metabolic turnover of membrane proteins. Thirdly, shed tumor antigens could reach the surface of a macrophage, perhaps aided by "natural" antibody, and this macrophage-bound, concentrated depot of anitgen, perhaps in a regional lymph node, could trigger as immune response.

We have been investigating lymphocyte triggering, using *in vitro* systems, with defined lymphocyte populations of either the thymus-derived, T cell family, or the bone marrow derived, B cell family. In the former case, the hydrocortisone resistant pool of mouse thymocytes serves as a convenient source consisting of > 99 per cent T cells and in the latter case, the spleens of congenitally athymic "nude" (nu/nu) mice represent populations of B cells mixed in with some macrophages and other cells, but free of T cells.

Consideration of the rules governing T or B lymphocyte activation are important to the cancer problem as each cell family plays a different role in tumor rejection. Activated T lymphocytes can kill tumor cells directly, and can also help B lymphocytes to form anti-tumor antibody. The progeny of B lymphocytes secrete various forms of antibody, which may have paradoxical and not fully understood effects on tumor cell growth. On the one hand, complement-fixing antibody can kill tumor cells, and, in the presence of very small amounts of antibody, normal B lymphocytes lacking specificity for tumor antigens, can engage in tumor cell killing (7). On the other hand, antibody can complex with tumor antigen, and such complexes can block cytotoxic killing of tumor cells by T lymphocytes (8).

In Vitro Models of T Cell activation

Our studies on *in vitro* T cell activation have been led by Dr. Hermann Wagner (reviewed in 9), and have used a two-stage tissue culture approach. In stage 1, mouse T lymphocytes are mixed with an antigen source and are cultured under rather special conditions for periods of 4 to 8 (usually 6 or 7) days. The antigen source consists of living cells bearing alloantigens or TATA, or of cell membrane fractions derived from such cells. If living cells are used, their potential for division is abrogated by prior treatment with the mitotic poison, mitomycin C. In stage 2, the cultured T cells are harvested, washed, and incubated for 3 to 6 hr with ^{51}Cr-labeled lymphoma or mastocytoma cells bearing the antigen (alloantigen or TATA) under study. The degree of activation of T lymphocytes, i.e. the relative number of "killer" cells, is measured by quantitating the amount of ^{51}Cr released (10).

The chief conclusions from this work were as follows:

1. In the system under study, only lymphocyte populations containing T cells could be activated to kill. Pure B lymphocyte populations could be

stimulated to divide by alloantigenic spleen cells, but could not develop into killer cells. T lymphocyte preparations uncontaminated by B lymphocytes could readily be activated.

2. Living mitomycin-treated cells were the best source of immunogenic stimulation. Cell membrane fragments were active, but much more weakly so. Lymphoid cells were more strongly immunogenic than fibroblasts. The quantitative degree of killing was much greater when allogenic cells acted as an antigen source rather than tumor cells bearing TATA.

3. Detectable activation of killer cells did not occur when macrophages were removed from the cell reaction mixture. In other words, the macrophage is an obligatory third partner with antigenic cell and responding T cell in the activation event. However, in this system, the macrophage appears to play no role in the final killing event itself.

4. Some evidence was obtained that the initiation of activation required collaboration between two types of T cell, but this has not yet been fully established.

5. Killing activity initiated *in vitro* was specific for cells bearing the antigens of immunogenic cells present during stage 1 of culture.

6. *In vitro* activated cells could be returned to *in vivo* situations, and could be shown to mediate allograft or tumor rejection.

7. The specificity of killer T cells appeared to reside in surface receptor molecules, which resemble 7S IgM but which possess sufficiently distinctive characteristics for us to regard them as being a separate immunoglobulin class, IgT (11, 12). IgT with binding specificity for tumor antigens or alloantigens has been isolated from the T cell surface. Excess amounts of cell membrane fragments containing alloantigens in soluble form can partially block killing by activated T cells.

8. Analysis of killer cells generated *in vivo* (13) or *in vitro* (9) suggested a progressive maturation sequence from rapidly-cycling large blast cell, to more mature killer cells which were smaller and had restricted potential for further division, and finally to non-dividing "end" cells. It was not excluded that these cells could live on after ceasing killer activity, reverting to a resting state but remaining ready to engage in a later cycle of blast transformation and clonal expansion.

We thus see that the sequence of cellular events leading to killer cell generation has many similarities to that leading to antibody formation.

In vitro models of B cell activation

In a parallel series of studies, Feldmann, Schrader and I have addressed ourselves to problems of B cell activation and tolerization. Here, the model antigens used were largely haptens coupled to pure proteins. We have reviewed this work elsewhere (11), and it remains only to state some general conclusions which incorporate recent ideas (14).

Activation of B lymphocytes appears to require two events or "signals" (15). "Signal 1" is union of lymphocyte surface receptor and antigenic determinant. "Signal 2" is some second, obligatory event, without which triggering will not occur, and which normally involves T cells and macrophages. The sequence which we propose is that T cell activation usually occurs first, possibly by an antigen reaching the macrophage surface. The activated T cell sheds from its surface IgT with specificity for the carrier portion of an antigen. Complexes of IgT and antigen are bound to the macrophage surface, there creating a depot of relatively concentrated antigen. B cells with receptors for those determinants of the antigen that are still free are attracted and held by multipoint binding of macrophage-associated antigen and B cell receptor immunoglobulin. The macrophage secretes a protein with "Signal 2" properties, which completes the triggering process. Under circumstances where "Signal 1" alone is presented to the cells for a sufficient period of time (perhaps a day or so) irreparable but poorly understood metabolic changes supervene which render the cell tolerant, i.e. unresponsive to normal triggering signals.

Differentiation of B lymphocytes.

So far, this analysis of T and B lymphocyte activation has neglected a fundamental parameter, namely the differentiation state of the cells under consideration. It is clear that both B and T cells are made continuously, even in adult animals, from more primitive precursor cells. These precursors lack Ig receptors. We have produced some evidence (16), by no means watertight, that there exists a stage during the maturation of the B cell when cells display a preferential tendency to become tolerized. In other words, antigen pulses given to cell populations all recently endowed with a receptor complement may cause tolerance rather than immunity. It is also evident that bone marrow lymphocytes are generated continuously and at a high rate in bone marrow, emerging from cell cycle as small lymphocytes which *lack* readily detectable Ig receptors. Over the next two to three days, a given lymphocyte appears to display more and more surface Ig, presumably reaching full reactivity to antigen at the end of this critical, non-mitotic maturation phase (17). Loss from the bone marrow and seeding to the periphery appears to be a random process, suggesting that some B cells undergo their final maturation outside the marrow. These results bear many similarities to cell differentiation in the thymus. In both situations, *antigen-independent* generation of lymphocytes occurs, with a morphologically visible maturation sequence ending in the production of non-dividing small lymphocytes. In both cases, maturation changes are reflected in measurable cell surface differences, and peripheral seeding of immunocompetent cells is the end result. Finally, as we have noted above, in both cases antigenic activation involves blast cell transformation in selected lymphocytes, with clonal proliferation and the generation of non-dividing immunological effector cells.

Summary

This paper has considered two broad areas of immunology that are intimately related. First, the concept of immunological surveillance has been reviewed. The existence in most tumors of TATA was affirmed, as was the existence of both T and B lymphocyte responses against many tumors. Evidence was briefly reviewed in man and experimental animals to show an increased tumor incidence in immunosuppressed states. The animal work suggests that this is most marked for virally induced malignancies. It is also possible that non-specific immunostimulation lowers tumor incidence, but here the evidence is not so compelling. The point was made that immunosuppression or immunodeficiencies of a congenital nature are rarely complete in either clinical or experimental situations, and thus the proportionate increases in cancer incidence which are seen may not reflect maximally on the efficacy of immune surveillance. On balance, it seems that this is a real and powerful mechanism.

The second area was intended to discuss lymphocyte differentiation in relation to antigenic activation, from the viewpoint of both thymus-derived (T) or bone marrow-derived (B) lymphocytes. Two cycles of what could be described as differentiation events occur in both cases. First, there is an antigen-independent and rapid generation of new lymphocytes in thymus or marrow. Maturation changes of definable nature can be observed at the lymphocyte surface. Secondly, there is a second cycle of cell multiplication and differentiation engendered by antigen. This involves selection of appropriate lymphocytes, blast transformation, clonal expansion and production of differentiated, specialized progeny. In each case, activation requires the participation of macrophages, and the possibility exists that these cells manufacture a chemical transmittor which aids triggering. In each case, a key early event in activation is union of antigenic determinant with a specific immunoglobulin molecule embedded in the lymphocyte membrane, but in neither case is this sufficient for activation. The exact nature of the necessary "second signal" for activation is still obscure, and is a subject of much current research. Stimulatory molecules from macrophages may be involved.

REFERENCES

1. Good, R. A. (1974). Immunodeficiency and Malignancy. In: "Host Environment Interactions in the Etiology of Cancer in Man," I.A.R.C. Monograph No. 5, (Lyon, France), Sir Richard Doll and I. Vodopija, eds., (in press).

2. Mäkelä, O. (1974). Influence of immunological reactions on carcinogenesis – an overview. In: "Host Environment Interactions in the Etiology of Cancer in Man," I.A.R.C. Monograph No. 5, (Lyon, France), Sir Richard Doll and I. Vodopija, eds. (in press).

3. Isliker, H. (1974). Discussion. In: "Host Environment Interactions in the Etiology of Cancer in Man," I.A.R.C. Monograph No. 5, (Lyon, France), Sir Richard Doll and I. Vodopija, Eds., (in press).

4. Nossal, G.J.V. (1974). The role of immunology and virology in Cancer Research in Man. In: "Host Environment Interactions in the Etiology of Cancer in Man," I.A.R.C. Monograph No. 5, (Lyon, France), Sir Richard Doll and I. Vodopija, eds. (in press).

5. Marshall, V. (1974). Premalignant and malignant skin tumors in immunosuppressed patients. *Transplant. Rev. 17*:272-275.

6. Möller, G. (1974). Keynote address. In: "Immunological Aspects of Neoplasia," 26th Annüal Symposium on Fundamental Cancer Research, M. D. Anderson Hospital and Tumor Institute, E. M. Hirsch and M. Schlamovitz, eds., (in press).

7. Perlmann, P. and G. Holm (1969). Cytotoxic effects of lymphoid cells *in vitro*. *Advan. Immunol. 11*:117-193.

8. Hellström, I. and K. E. Hellström (1969). Studies on cellular immunity and its serum-mediated inhibition in Moloney-virus-induced mouse sarcomas. *Internat. J. Cancer 4*:587-600.

9. Wagner H. and G.J.V. Nossal (1973). Transplant Rev. (in press).

10. Cerottini, J-C and K. T. Brunner (1974). Cell-mediated cytotoxicity, allograft rejection, and tumor immunity. *Advan. Immunol. 18*:67-132.

11. Feldmann, M. and G.J.V. Nossal (1972). Tolerance, enhancement and the regulation of interactions between T cells, B cells and macrophages. *Transplant. Rev. 13*:3-34.

12. Marchalonis, J. J. and R. E. Cone (1973). Biochemical and biological characteristics of lymphocyte surface immunoglobulin. *Transplant. Rev. 14*:3-49.

13. Sprent, J. and J. F. Miller (1971). Activation of thymus cells by histocompatibility antigens. *Nature New Biol. 234*:195-198.

14. Schrader, J. W. (1973). Specific activation of the bone marrow-derived lymphocyte by antigen presented in a non-multivalent form. *J. Exp. Med. 137*:844-849.

15. Cohn, M. (1971). The take-home lesson. *Ann. N.Y. Acad. Sci. 190*:529-584.

16. Nossal, G.J.V. and B. Pike (1974). In: "Immunological Aspects of Neoplasia," 26th Annual Symposium on Fundamental Cancer Research, M.D. Anderson Hospital and Tumor Institute, E. M. Hirsch and M. Schlamovitz, eds., (in press).

17. Osmond, D. O. and G.J.V. Nossal (1974). Differentiation of lymphocytes in mouse bone marrow: II Kinetics maturation and renewal of antiglobulin binding cells studied by double labeling. *Cellular Immunol. 13*: (in press).

Subject Index

A 4
B 5
C 6
D 7
E 8
F 9
G 0
H 1
I 2
J 3